CMDA IS-95
for Cellular and PCS

CMDA IS-95
for Cellular and PCS

Lawrence Harte

McGraw-Hill

New York   San Francisco   Washington, D.C.   Auckland   Bogotá
Caracas   Lisbon   London   Madrid   Mexico City   Milan
Montreal   New Delhi   San Juan   Singapore
Sydney   Tokyo   Toronto

McGraw-Hill

*A Division of The **McGraw·Hill** Companies*

1 2 3 4 5 6 7 8 9 0 DOC / DOC 9 0 9 8 7 6 5 4 3 2 1 0 9

ISBN 0-07-027070-8

The sponsoring editor for this book was Steve Chapman, and the production supervisor was Pamela Pelton.

Printed and bound by R.R. Donnelley & Sons Company.

McGraw-Hill books are available at special quantity discounts to use as premiums and sales promotions, or for use in corporate training programs. For more information, please write to the Director of Special Sales, McGraw-Hill, 11 West 19th Street, New York, NY 10011. Or contact your local bookstore.

 This book is printed on recycled, acid-free paper containing a minimum of 50% recycled de-inked fiber.

Acknowledgment

We thank the many gifted people who gave their technical and emotional support for the creation of this book. In many cases, published sources were not available on this subject area. Experts from manufacturers, service providers, trade associations and other telecommunications related companies gave their personal precious time to help us and for this we sincerely thank and respect them.

We thank the many manufacturing experts that helped us validate technology information including: Al Fisher from Anritsu, Yas Mochizuki with Casio, Pat Kennedy at CellPort Laboratories, Shahin Hatamian with Denso Wireless Communications, Gabriel Hilevitz of DSPC Israel Ltd., Eric Stasik from Ericsson Radio Systems AB, Osmo Hautanen, at Formus Communications, Richard Holder with HebCom, Beth Eurotas of Hewlett Packard, Bob Sarwacinski from Insight Technologies, Mat Kirimura at Japan Radio Company, Ronald Koppel, Vijay Garg, Ph.D. with Lucent Technologies, Greg Foss, Mark Worthey, Sherri Haupert and YS Cho of Maxon, Brian Walker, Joshua Kiem and Robert Dunnigan from Motorola, Kim Kennedy at NEC America, Inc., Denise Borel with Nortel Networks, Mike Wise of Oki Telecom-GA, Bob Roth, Christine Trimble, Gill Harneet, Joanne Coleman, Kevin Kelly and Michelle French from Qualcomm, Ram Velidi, Ph.D. at Raytheon TI Systems, Inc., Kurt Siem, Rich Conlon with Repeater Technologies, Dawn McLain of Samsung, Dan Fowler at SCALA and Stuart Creed with Tektronix.

Special thanks to the gifted wireless carrier professionals who helped us understand the new services of CDMA technology and how network deployment is occurring. These people include Kei Foo from Alltel, Cory Linquist with General Wireless, Stephen Rainbolt of GTE Communications Corporation, Geoff Livingston at IPNet Solutions, Allen Salmasi with NextWave Telecam Inc., John Nation from Peachtree Telecom International, inc, Ken Geisheimer, Limond Grindstaff and Scott Akrie with PrimeCo Personal Communications, Michael Ha of Shinsegi Telecomm Inc., Donald Prophete from Sprint PCS and Rafael Andrade at TEC Cellular.

Thanks to experts at research, consulting and design firms for providing us with industry insight and hard to obtain market and technology information. These people include Syang-Myau(Kevin Hwang,Ph.D., Cadence Design Systems Inc., Leila Ribeiro, Leonhard Korowajczuk and Oscar Miranda of Celplan Technologies Inc., Herschel Shosteck and Jane Zweig from Herschel Shosteck Associates, Simon Hsiao with LCC Inc., Masud Tarafder, Tatiana Ruiz and Thomas Tran of MLJ, Elliott Hamilton with Strategis Group and Konny Zsigo of Zsigo Wireless.

We are very grateful to the leading associations and education experts who include Christine Bock and Perry LaForge at the CDMA Development Group (CDG), Barry Schiller with the ITRE, Bradley Eubank from PCIA and Dr. Keith Townsend at NC State University.

Our gratitude also goes to the finance experts at leading brokerage firms for their financial analysis of wireless systems. These experts include: Alex Cena and Karen Nielsen at Soloman Bros Smith Barney, Jeff Hines with NatWest Securities and Jeffrey Schlesinger from Warburg Dillon Read.

Special thanks to the people who assisted with the production of this book including: Judi Rourke O'Briant (project manager), James Harte (researcher), Nancy Campbell (graphics and layout). And thanks to our other supporters including Michael Shafer and Jerry Eatman.

About the Authors

Lawrence Harte is the president of APDG, a provider of expert information to the telecommunications market. Mr. Harte has over 19 years of experience in the electronics industry including company leadership, product management, development, marketing, design, and testing of telecommunications (cellular), radar, and microwave systems. He has been issued patents relating to cellular technology and authored over 75 articles on related subjects. Mr. Harte earned his Bachelors degree from University of the State of New York and an executive MBA at Wake Forest University. During the TDMA cellular standard development, Mr. Harte served as an editor for the Telecommunications Industries Association (TIA) TR45.3, the digital cellular standards committee.

Mr. Koenig is a Senior Engineering Instructor at QUAL-COMM Incorporated. Mr. Koenig has over 10 years experience in advanced telecommunication networks through training and support programs. These programs are targeted for marketing and sales, hardware and software design engineers, and network planners. Previously, Mr. Koenig was a communications specialist with the military where he gained electronic communication systems experience. He received his education and experience at high level communication schools in the military and holds a degree in Education.

Daniel McLaughlin is a radio frequency engineer for Sprint PCS. At Sprint PCS he is responsible for CDMA system design including placement of cell sites, site candidate selection and evaluation, drive tests, individual site configuration, and both single site and system optimization. Mr. McLaughlin's experience includes initial design, implementation, performance testing and optimization phases of the deployment process and conducting intermodulation studies. Mr. McLaughlin also ensures compliance with FAA, FCC, NEPA and other regulatory requirements. Prior to joining Sprint PCS, Mr. McLaughlin worked for IBM and was involved with wireless data software and equipment. He has extensive experience with software tools that test and operate telecommunications equipment. Mr. McLaughlin has also developed and instructed various courses on communication and electronics technology. Mr. McLaughlin is a graduate of North Carolina State University and holds bachelors degrees in electrical engineering and computer engineering.

Mr. Roman Kikta is the Director Product Concepts, Applications & Technologies - Americas at Nokia Mobile Phones. Mr. Kikta is responsible for the identification, development and implementation of all new product concepts, applications, value-added services, business ventures and strategic alliances and partnerships in North and South America. He is also a member of the global Concept Creation Design Group; SocialWare and Futures Group whose focus is to track world social trends and provide direction for future strategies, products and services. Mr. Kikta previously served as Nokia Director of Marketing - PCS, where he was respon-  sible for directing the development of Personal Communication Services (PCS) product programs, marketing strategies, distribution plans, new business development, and coordinating all marketing, sales, and engineering efforts that resulted in a successful product launch and the first deployment of PCS in the U.S. During his 16 years in the cellular and PCS wireless telecommunications industry, he has held senior level product planning & development, sales, marketing and market development positions with leading cellular and wireless manufacturers including Panasonic, GoldStar, and OKI Telecom. Mr. Kikta has written several articles on various cellular and PCS telecommunications issues for industry publications and has served as a guest speaker at several telecommunication industry conferences and trade shows in the U.S. and internationally. He is a graduate of Rutgers University in New Jersey.

Dedication

"I dedicate this book to my parents: Virginia and Lawrence M. Harte, my children: Danielle Elizabeth and Lawrence William and the rest of my loving family."
  Lawrence

"I dedicate this book to my parents: Melvin and Leanna Koenig who encouraged me to pursue my career."
  Morris

"I dedicate this book to my parents: Robert and Kathy McLaughlin, my brother Jarrett McLaughlin, and the rest of my loving family"
  Daniel

"With special appreciation to Mal Gurian who welcomed me into the wireless world back in 1983."
  Roman

Foreword

Since the first commercial CDMA telephones were introduced into the marketplace in 1994, the demand for CDMA digital telephones and service continues to grow at over 120% per year. At the end of 1998, there were 304.9 million mobile subscribers throughout the world. Of these, 99.8 million were in Europe; 75.2 million in North America; 39 million in Japan and 60.1 million in the rest of Asia; 20.8 million in South America and 10 million in the Middle East region.

Within the next five years, it is projected that over half (some predictions show up to 63%) of all Americans will carry a wireless phone. Of these wireless phones, approximately half are estimated to use IS-95 CDMA wireless access technology. The available features and services on wireless networks are changing rapidly. The 21st century will see new lifestyles that are enabled by the new capabilities of digital wireless communication.

New digital products & services are re-invigorating the market, and predictions show that there will be over 700 million users by the end of 2003 and over 1 Billion wireless telephone users world-wide by the year 2005.

The first commercial communication system that used digital transmission was via wires in 1963. This digital system allowed several users to share the limited resource of copper cable via digital transmission. It wasn't until the mid 1980's that the efficient use of digital transmission could be applied to wireless communication.

There were several key innovations that made digital radio transmission for mobile phones feasible. These include shared radio access technology, high speed digital signal processing, and industry standards.

In the early days of radio, each radio user would share a single radio channel. This evolved to sharing radio channels through time division. The latest innovation in digital wireless technology is sharing the code division.

Digital radio transmission signal processing that is fairly complex and is not practical to implement using off-the-shelf electronics components. Advanced digital signal processing was necessary to allow the technology to be used in mobile radios.

Finally, industry standards were necessary to allow common technology to be shared among many manufacturers. This was necessary to ensure the marketplace would be big enough to justify the technology development for advanced digital cellular products. Over 100 manufacturers participated in the development of the code division multiple access (CDMA) industry standard.

Furthermore, because of a unique convergence of factors, including the market's insatiable demand for Internet access, it is believed that the market for wireless data services, while relatively small today, is on the verge of experiencing explosive growth similar to what has been seen on wireline networks. According to the Strategis Group, an independent market research group, the number of wireless email and wireless Internet users will grow from only 2 million in 1998 to almost 30 million by 2003, a 1500% increase in only five years. Another forecast by Ovum, an international research and consulting firm, projects the number of wireless data subscribers in North America to grow to 72 million by 2007.

This book arrives at an opportune time in the wireless industry. The amount of capital and new technology that has been invested in wireless technology is immense and growing. While digital wireless technology has proven its success, the deployment of new capabilities to emerging markets requires an understanding that balances between market needs and technology availability. I believe this book is unique in describing the existing and new services digital wireless can provide, how IS-95 CDMA can provide these services and the financial considerations for its deployment.

Allen Salmasi
Chairman of the Board, President and CEO, NextWave Telecom Inc.

Table of Contents

Preface

Since the first commercial CDMA telephones were introduced into the marketplace in 1995, the market demand for CDMA digital telephones and service continues to grow at over 120% per year. According to Strategis Group research, there will be over 450 million cellular and PCS telephones in use worldwide by the year 2000 and a large percentage of these will have CDMA digital transmission capability.

There have been many claims of advanced features and economic benefits to digital cellular technologies. This book provides a semi-technical understanding of the IS-95 CDMA system, its economics and the advanced services it can provide. IS-95 CDMA technology has unique advantages and limitations that offer important choices for managers, technicians, and others involved with CDMA wireless telephones and systems. IS-95 CDMA, Technology, Economics and Services provides a description of CDMA technology, shows the economic benefits, and provides references for suppliers and industry specifications.

This book explains CDMA technology and its services by using over 120 illustrations and tables. More than 100 industry experts have helped create and review the technical content of this book. Industry terminology is explained and chapters in this book are organized to help find the needed detailed information quickly. These chapters are divided to cover specific parts or applications of CDMA technology and may be read either consecutively or individually.

Chapter 1. Provides a basic introduction to wireless technologies including cellular, wireless office, cordless, and PCS. Advanced wireless messaging services such as advertising, imaging, and monitoring are identified and explained. This chapter is an excellent introduction for newcomers to wireless technology.

Chapter 2. Covers analog cellular technology and the basic operation of cellular systems. This chapter includes an overview of analog cellular, radio channel structure, signaling and call processing.

Chapter 3. This chapter provides an explanation of CDMA technology. It includes the radio channel structure, signaling messages, system parameters; and key features such as power control, soft handoff, frequency diversity, time diversity/rake reception, and variable rate speech coding.

Chapter 4. Describes CDMA mobile telephone operation, design, and options. The descriptions cover the radio section, baseband signal processing, power supply, accessories and the call processing to make all the sections work together. Design options including DSP and ASIC tradeoffs are covered along with multiple frequency superphones.

Chapter 5. Explains CDMA networks including and functional section descriptions. Base stations and switching system equipment is described. Intra-network signaling and inter-network connections are explained along with key design and implementation options.

Chapter 6. This chapter provides an overview of the key testing requirements for CDMA mobile telephones, network equipment and system field testing for CDMA systems.

Chapter 7. This chapter explains the marketing and economics factors that can significantly affect the roll out of CDMA technology. Included are market demand estimates, mobile telephone costs, system equipment costs, operational costs, and key marketing considerations.

Chapter 8. Identifies and analyzes the key mobile features and services that make CDMA such an attractive technology. These include extended battery life, data services, enhanced voice quality and others.

Chapter 9. Describes features and services that can be provided by CDMA systems. These include short message service, circuit and packet data switching and other call processing features.

Chapter 10. Describes future advances in CDMA technology. Some of the advances that are discussed include wireless intelligent networks, universal wireless access protocol, spatial division multiple access (SDMA), satellite cellular systems and 3rd generation cellular systems..

Appendix 1 defines the common acronyms that are associated with CDMA and wireless technology. Appendix 2 contains a listing of the CDMA equipment suppliers. Appendix 3 contains a world list of CDMA systems.

Chapter 1

Introduction to IS-95 CDMA

IS-95 CDMA (commonly called "CDMA") is a digital cellular radio system that is used in over 35 countries throughout the world. The IS-95 digital cellar system optionally combines analog cellular communications network with digital CDMA radio technology. In 1998, there were over 23 million CDMA phones in use [1] and projections show that by the year 2002, there will be over 106 million CDMA phones [2].

The CDMA network provides for mobile voice communication as well as many new advanced services like mobile fax and text messaging. This book explains in both simple language and in detail how CDMA and its applications work. The authors hope that it will be useful to those new to telecommunications as well as those somewhat experienced in the field.

This book is intended for persons having a general familiarity with cellular and PCS networks, and a particular interest in CDMA and PCS-1900 technology. The technical background level expected from the reader is minimal.

What is CDMA

CDMA is the abbreviation for Code Division Multiple Access communication. CDMA is a form of spread spectrum communications. Spread spectrum communications is the transmission of a radio signal over a radio channel that is much wider than necessary to transmit the original information signal. Because the signal is spread over a very wide bandwidth, interference from other users within that bandwidth is minimal. This allows multiple users to share the radio channel at the same time.

There are two basic types of spread spectrum communications: frequency hopping and code division. Frequency hopping multiple access (FHMA) is an access technology that allows mobile radios to share radio channels by transmitting for brief periods of time on a single radio channel frequency and then hopping to other radio channel frequencies to continue transmission. Each mobile radio is assigned a particular hopping pattern and collisions that occur randomly occur and only cause a loss of small amounts of data that may be fixed through error detection and correction methods. CDMA technology is called a wideband spread spectrum system as compared to earlier narrowband wireless systems. A wideband spread spectrum system spreads the radio signal over a frequency bandwidth that is much wider than is necessary to transfer the information signal (typically voice). By spreading the radio signal over a wide frequency bandwidth, this reduces the interference to other users operating in or near the radio channel bandwidth. Code division multiple access (CDMA) allows multiple users to share a single radio channel frequency at the same time by assigning a unique code sequence to each mobile radio.

History of CDMA

Development of CDMA technology originated in the United States in 1989 as a result of the CTIA next cellular generation technology requirements. In September 1988 the Cellular Telecommunications Industry Association (CTIA) laid out the User Performance Requirements (UPR) for the next generation of wireless service. The requirements were for a digital technology that would include:

- Tenfold increase over analog system capacity
- Ability to introduce new features
- Higher voice quality
- Voice and Data Privacy
- Ease of transition and compatibility with existing analog system

In 1989 the Telecommunications Industry Association (TIA) adopted time division multiple access technology (TDMA) as the radio interface standard. With the support of infrastructure equipment, MS manufacturers, and carriers, a company called QUALCOMM developed a CDMA system compliant with the CTIA requirements. In December 1991 QUALCOMM along with participating carriers and manufacturers, presented the results of the CDMA system field trials. In 1992 the CTIA Board of Directors adopted a resolution requesting TIA to prepare structurally to accept contributions regarding wideband systems. TIA unanimously adopted the motion and recommended that the TR45 Committee address standardization activities regarding wideband spread spectrum digital technologies. In July 1993 TIA voted on and accepted IS-95 as the CDMA air interface standard (radio specifications). CDMA systems based on the IS-95 standard and related specifications are referred to as cdmaOne? systems.

The first commercial network began operation in Hong Kong in 1995. Since then commercial service has begun in both cellular and PCS bands throughout the world. CDMA is the fasted growing technology in wireless communications. Figure 1.1 illustrates the time between the deployment of the first Advanced MS Phone Service (AMPS) network, CTIA's UPR announcement and acceptance of TDMA, and IS-95 CDMA.

AMPS Deployment	New CTIA Requirements	TDMA Specifications	CDMA Specifications	First CDMA System Korea	23 Million CDMA Users
1983	1983	1989	1993	1995	1998

Figure 1.1, Wireless Development Timeline

Forms of CDMA

cdmaOne™ is a trademark of the CDMA Development Group and refers to the family of standards that define the CDMA technology. IS-95 specifies the AMPS and CDMA operation in the 800 MHz cellular frequency band. The IS-95 specification is a dual mode technical standard that incorporates both IS-553 AMPS and CDMA functions into one standard.

This standard defines the mobile station (MS) and base station (BS) compatibility requirements for CDMA and analog operational modes. A revision of the standard (IS-95A) was voted on and accepted in 1995. This standard adds additional features to the original CDMA system.

New features and capabilities became available but were not available for inclusion into IS-95A. The features and capabilities were for a new voice encoder/decoder and extended messages. The extended messages and features are defined in TSB-74.

In 1998 revisions to the standard were accepted. The new standard is called TIA/EIA-95. The standard now incorporates IS-95A, TSB-74, and ANSI J-STD-008. The analog portion of the standard has been removed and referenced where needed. In addition to combining everything into one standard, there are several new enhancements and corrections. Among the new features are enhancements to the access process, Traffic Channel handoff process, support for medium data rate services, position location, and subscriber addressing.

TIA/EIA-95 is the minimum compatibility requirements for the MS and BS. The standard defines the modulation scheme for the six code channels, power control, call processing, handoffs, and registration techniques. A CDMA network includes the same basic subsystems as other wireless systems, including a switching network, controller, Base Station (BS), and Mobile Station (MS).

Figure 1.2 shows how the CDMA standard was developed. This diagram shows that the features, services and requirements were created by standards organizations using many proven technologies. This is why CDMA is called a "second generation" cellular system.

The different revisions of CDMA technology can make use of the same digital radio channel structure that set up handover and end the connection. These digital signals are sometimes referred to as the "base band" waveforms, and the part of the hardware that processes these is sometimes called the base band hardware. While the base band hardware can be absolutely identical in mobile stations and handsets that use CDMA standards, they can operate on different frequency bands. Only the radio portion of the handsets and base stations are substantially different in hardware, and in some cases the differences are very minor indeed. Thus, the CDMA standard can be implemented at 800 MHz in one location, and then at 1900 MHz in the another location. Even though the technology is the same, because the frequency of the network determines the frequency of the hardware required, a single mode 800 MHz CDMA handset will not function in an 1900 MHz CDMA network. However, some manufacturers produce handsets that can operate on either fre-

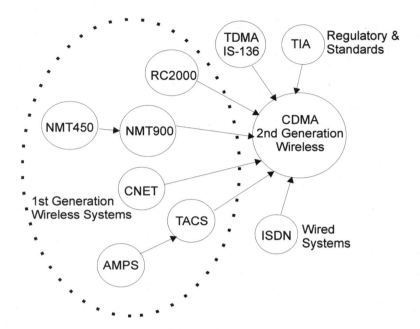

Figure 1.2, Development of CDMA Standard

quency band (dual band handsets).

All CDMA networks use a specific radio frequency band for signals from the base transmitter to the mobile receiver (called the forward or downlink channel) and a second distinct radio frequency band for the signals from the mobile transmitter to the base receiver (called the reverse or uplink channel).

In CDMA, the separation "on the radio dial" between adjacent radio carrier frequencies is 1.25 MHz (megahertz). A pair of radio carrier frequencies is loosely referred to (matching uplink and downlink) as "a frequency." In both documentation and in control signals sent between the base and mobile stations, each distinct matching carrier frequency pair is designated by an identification number such as 1, 2, 3 and so forth. To clarify the terminology, CDMA documents consistently distinguish between the words "carrier" and "channel." In CDMA, a carrier frequency is divided by means of codes into 64 individual channels. Each channel carries the information related to a separate and distinct conversation in digitally coded form. In some cases, an individually coded channel carries signals related to the beginning of a connection. Confusion sometimes arises when discussing older analog frequency division multiplex (FDM) systems, in which each pair of carrier frequencies can carry only one conversation. In FDM systems, a channel is synonymous with a carrier. In spread spectrum systems such as CDMA, one carrier carries several channels.

800 MHz CDMA

CDMA technology can be used in existing cellular frequency bands and the new personal communications service (PCS) frequency band. When used in the cellular system, CDMA operates in same radio spectrum allocation for cellular systems. It maintains a separation of forward and reverse channels in cellular band is 45 MHz. The MS transmit frequency band is 824-849 MHz. The BS transmit frequency band is 869-894 MHz.

In the CDMA cellular network, some radio carrier frequencies are defined for CDMA use. Not all of these frequencies are used for CDMA transmission. The FCC requires that analog radio transmission (AMPS) continue to operate. In the late 1990's, almost all the 800 MHz CDMA phones that were produced were capable of operating on CDMA or AMPS radio channels. This is called dual mode operation.

1900 MHz CDMA (PCS)

Some CDMA systems operate in the new Personal Communications System (PCS) frequency bands. PCS is primarily available in North America on the 1900 MHz frequency band, where it is called PCS-1900. Because of the commonality of the base band signals, some manufacturers make dual band CDMA handsets that can operate on both the 800 MHz and 1.9 GHz bands. When the base CDMA networks in the same region operating on these two bands are properly linked, it is possible for a subscriber to obtain service from either or both such systems if their handset is equipped for dual frequency band operation.

When used in the PCS frequency band, CDMA is specified for operation under the ANSI J-STD-008 specification. This standard is an up-banded version of IS-95A and TSB-74 but without the analog compatibility requirements. CDMA operation complies with the frequency structure of the PCS band. The separation of forward and reverse channels in the 1800 MHz PCS band is 80 MHz. The MS transmit frequency band for is 1850-1909 MHz. The BS transmit frequency band for is 1930-1989 MHz.

CDMA Parts

A CDMA network is comprised of several major portions: a mobile radio part, subscriber information part, a radio network, a switching system and network intelligence (primarily databases). Figure 1.3 shows a basic CDMA network. The mobile phone is called a mobile station. There are several types of mobile stations in CDMA. High-power mobile phones can be used in vehicles and people typically carry low-power mobile phones (handhelds). Mobile stations communicate with nearby radio towers called base stations. Base stations convert the radio signal for communication to a switching system. The switching system connects calls to other mobile stations or routes the call to the public telephone network. The switching system is connected to several databases that hold customer information. These databases include phone numbers, electronic serial numbers and authorized feature lists (features the customer has subscribed to).

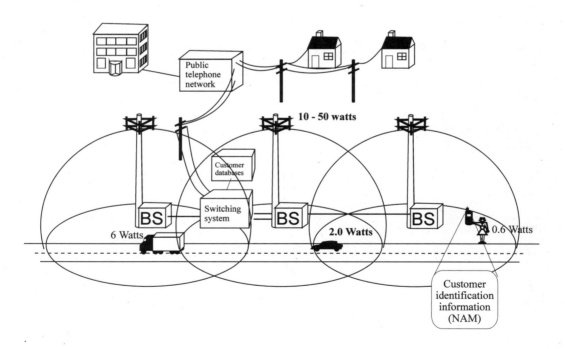

Figure 1.3, Basic CDMA Network

Mobile Station

The mobile station (MS) is the subscriber's interface with the CDMA network. In the 800 MHz CDMA network, both hand-held MS units having a low-power radio transmitter and vehicle-mounted MS units are permitted. For PCS versions of CDMA, a low-power MS transmitter is the rule and handsets are only nominally used. Although, some handset manufacturers provide specific adapters for vehicle use which give coupling to an antenna outside the vehicle and, in some cases, extra mobile transmit power as well.

Most handsets use a rechargeable battery (called a secondary cell) for power. Rechargeable Nickel Cadmium or Lithium cells are typical. Some models can use disposable (called primary cells) batteries as a temporary source of power as well. Primary cells are readily available as they are widely used for flashlights, toys, etc. Most handsets come with a recharging device for the rechargeable cells, operating from either alternating current wall outlet (mains) power in a building, or a convenience outlet in a vehicle, or both. Several aspects of the CDMA design, explained in later chapters, help to give long stand-by time and long talk time with minimal battery drain.

All CDMA handsets built to date have voice capability, with a microphone and an earphone to speak and listen. The user does not need to operate a push-button to change from talk to listen, as in some older radio systems. The speech is digitally coded for transmission over the radio link. The digital speech compression process is performed by a codec. The acronym "codec" is a contraction of the first part of the two words: coder and decoder.

There are two types of digital speech coding used in the CDMA system. The original speech coding systems used 8 kbps speech coding and the later versions use 13 kbps speech coding. The 13 kbps speech coder was developed to offer higher speech quality.

Unlike other digital cellular systems, the CDMA system uses variable rate speech coding. As the speech activity varies (e.g. talk and silence), the data rate of the speech coder changes. This results in a data transfer rate that is lower for compressed speech signals (typically 40% of the designed data rate).

CDMA handsets also have an array of push buttons for dialing and for originating and answering calls. The numeric dial buttons can also be used to produce "touch tone" or dual tone multi-frequency (DTMF) tones while connected to a voice line, so that a user can operate such devices as remote control answering machines or the like.

Some CDMA handsets have electrical connectors for use with an external fax machine or data terminal. The basic data transmission rate for a CDMA radio channel is 9.6 kbps. High-speed data services have been developed to allow several CDMA communication channels to be combined to achieve data rates of over 56 kbps.

Each handset contains a radio receiver and transmitter (the combination is sometimes called a transceiver), and a radio antenna. The radio and other parts of the handset are controlled by a microprocessor. The technical details of these parts of the set will be described in a later chapter.

Base Station Subsystem (BSS)

The radio parts of the CDMA network equipment are contained within the Base Station Subsystem (BSS). The Base Station Subsystem is divided into two main parts: the Base Transceiver Station (BTS) and the Base Station Controller (BSC). The BTS comprises several base radio transceivers. Each transceiver consists of a transmitter and a receiver which has a duplicated "front end" to match up with the two receiving antennas used in the base antenna assembly. The BSC comprises a control computer (typically a microprocessor central processing unit with memory), data communication facilities, and multiplexing and de-multiplexing equipment. The BSC can control the radio power levels of the various transceivers in the BTS, and also can autonomously control the mobile stations' radio transmitter power levels as well. The BSC passes certain types of control messages between the BTS and the Mobile Switching Center, and handles certain types of control messages itself under appropriate conditions. A single BSC can control several BTS radio equipment transmitters. The BSC can be located in a base station or at another remote site.

Figure 1.4 shows a basic diagram of a CDMA base station sub-system. The BTS consists of transmitters, receivers, antenna assembly, power supplies and test circuits. In this diagram, the BSC is located at the base station. Each transmitter operates on a different radio carrier frequency. Each radio carrier is divided into time slots and frames. For typical CDMA handsets (called full rate), this allows a maximum of 64 different communication channels. Because some of these communication channels are dedicated as control channels and some are simultaneously assigned to mobile radios that are transferring calls to other cell sites, up to 40 users can share a single radio carrier channel.

One (or more) radio carrier codes (communication channels on a single RF carrier) is used as a control channel. The control channel coordinates mobile station alerting and access to the CDMA network. A special version of control channel called the paging channel sends out the paging messages to alert mobile radios of an incoming call.

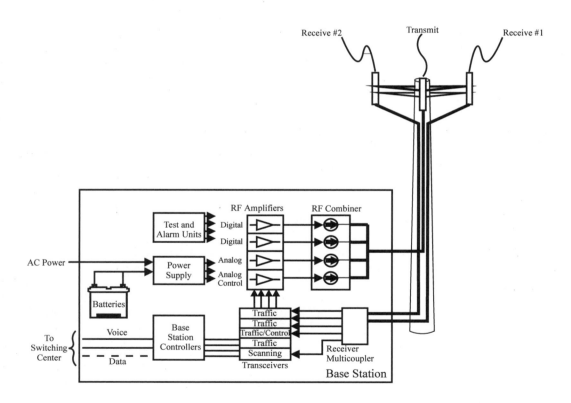

Figure 1.4, CDMA Base Station Subsystem

Network and Switching Subsystem (NSS)

The CDMA network requires a switching network and intelligence to interconnect calls between mobile phones and the public telephone network. The central switch of a CDMA installation is called a Mobile-service Switching Center (MSC). In earlier documents, the word "service" was omitted, which gave some people the incorrect impression that the MSC was itself mobile or capable of motion while in service. The MSC is, in all modern CDMA networks, an electronic digital telephone switch with digital multi-channel (trunk-type) telephone line inputs and outputs. Trunks are telephone channels that connect between one switch and another. Different subscribers in each successive connection use trunk channels. In an MSC, some of the trunks connect the MSC to a BSC, while other trunks connect the MSC to the PSTN switches.

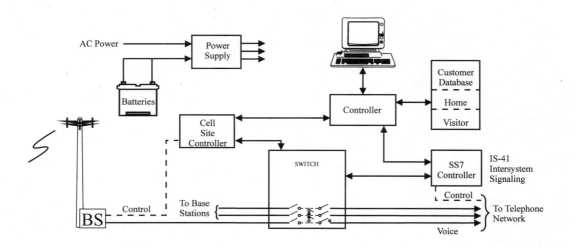

Figure 1.5, Mobile Switching Center

Figure 1.5 shows the basic building blocks for the MSC. The MSC consists of a switching center, power supplies, and alarm monitoring equipment. The switch is also connected to customer databases that may be located at the MSC or located at a remote site. In this diagram, the switch allows connection between each base station and the public telephone network. While the diagram shows physical switches, most modern switching system use electronic switching systems (ESS). ESS systems use a process called time slot interchange (TSI) to connect incoming and outgoing digital lines together through the use of temporary memory locations. The TSI system uses a computer to control the assignment of these temporary locations so that a portion of an incoming line can be stored in temporary memory and retrieved for insertion to an outgoing line.

There are many network processing centers and databases used in a CDMA network to check authorization for service and process call features. The most utilized network database parts store and process home customer subscriber lists, hold temporary customer information, validate equipment identity information (authentication), manage the fraudulent equipment identity list, and store and forward messages.

A Home Location Register (HLR) database holds the detailed subscriber service subscription information. This database can be located with the MSC, or it may be at a distant location. In some implementations, multiple MSCs share the same HLR. The HLR holds a user profile that indicates if a particular user subscribes to services such as call forwarding, call waiting, etc. The HLR also stores information about the present location of its subscribers who are presently visiting in the radio service area of another MSC, and indicates whether or not they have arranged to receive calls there.

A Visitor Location Register (VLR) database holds temporary information about active subscribers that are operating within the control of that particular MSC. This includes both visiting and active local subscriber data. The word "visited" is somewhat misleading since the data here is not restricted to visitors. The data to a large extent is a copy of the corresponding subscriber data taken from the HLR. The VLR is usually built into the MSC. In some implementations, HLR and VLR are the same physical data base, with records active in the VLR specially/temporarily marked as required, rather than copied from one database to another.

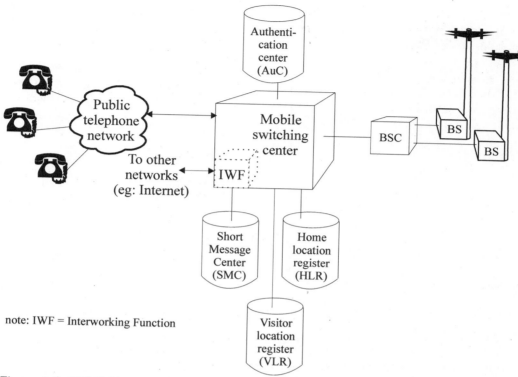

Figure 1.6, CDMA Network Parts

A Short Message Control center (SMCC) stores and forwards short messages to and from the CDMA network. An interworking function (IWF) is used to process and adapt information between dissimilar types of network systems. Figure 1.6 shows the basic parts of a CDMA network. In this diagram, several databases are inter-connected to each other and to a MSC.

Operation and Maintenance Subsystem

The Operation and Maintenance Subsystem (OMS) includes alarms and monitoring equipment to help a network operator run, diagnose, and repair a communications network. This includes administrative subscriber and mobile station management systems such as billing, accounting and statistics. It also includes access security, system performance monitoring, system design changes and maintenance. There are several databases that support the operation and maintenance subsystem.

These include a subscriber validation processing center and a fraudulent equipment database. A network operator including customer care, marketing and billing database may use many other databases.

An authentication center (AuC) is a database and processing center that is used to validate the identity of mobile stations. The AuC processes secret information (electronic keys) that is transferred between a mobile station and the CDMA network.

Industry Standard and User Groups

This section summarizes the various industry standard and user groups that are active in production of standards and information relating to CDMA.

CDMA Development Group (CDG)

The CDMA development group (CDG) represents manufacturers and carriers that provide CDMA related products or services. Information for this user group can be found at www.cdg.org.

Telecommunications Industry Association (TIA)

The Telecommunications Industry Association (TIA) is a telecommunications industry trade association represents the manufacturers of telecommunications equipment. The TIA is responsible for overseeing the production of wireless specifications in the United States.

European Telecommunications Standards Institute (ETSI)

The European Telecommunications Standards Institute (ETSI) is a source of some CDMA technical standards. These standards are available in both paper and CD-ROM form.

Telecommunications Planning Group (T1.P1)

This US based standards group is a joint sponsor coming from the Alliance for Telecommunications Industry Solutions (ATIS) and the document source of the J-007 standard for North American PCS-1900 standards.

Basic Operation

There are many processes a mobile station must perform to operate in a CDMA network. The basic call processing operation of a mobile station includes initialization, system access, paging and handover.

Mobile Station Initialization

When a CDMA handset is first powered on in a CDMA network, it begins an initialization process prior to accessing the system. The initialization process involves finding a suitable radio carrier channel and capturing system information that allows the mobile station to access the system.

When seeking a radio carrier signal with strong signal strength, it will typically find several frequencies. Having scanned for radio carrier channels, the MS then goes back and examines each frequency beginning with the strongest signals. It is seeking a radio carrier channel that contains a control channel. These channels are identified by pilot channel codes. After it has found a control channel, the MS begins to receive and store certain system broadcast information. This broadcast information includes data that allows the MS to access to the system.

System information is continuously sent by the system to allow mobile radios to determine how they access the system and what special system features are offered. This information includes system identification, the maximum access power level at which the MS should transmit when requesting service, locations of paging and messaging channels, and other information that coordinates access to the CDMA network.

Every installed CDMA base service area has a unique system identification code (SIC), a number which it broadcasts periodically and which identifies it distinctly from other system operators in the same city or anywhere else in the world. Each base station also broadcasts a message that tells mobile stations in that cell how much RF power they should use when initially transmitting a signal to the base station. Small cells request low power and large cells request high power. The mobile set also has the SIC of its own home system that was previously stored in its internal memory. The mobile set is also able to measure the radio signal strength of each such pilot channel. Given all this information, the mobile station chooses the "best" pilot channel and sends an identifying message. The best frequency is one that has the home SIC stored in the handset memory. If the MS is in its own home city, this rule will cause it to temporarily ignore the pilot channels of other system operators in that city in favor of its own home system. If that is not available in the vicinity (which would happen if the MS were roaming to another city at this time) the MS treats all SICs equally. The MS then chooses the pilot channel with the smallest cell so it can use the lowest transmit power, and transmits certain signals which identify the MS to the base system. The details of these signals will be explained in a later chapter.

Incidentally, when a MS has found pilot channels that are in use in the various cells in a city, it stores these carrier frequency numbers in the handset temporary memory. Then, when the MS power is turned on and such a list is available in its memory, the MS can be ready for service much sooner, because it only needs to scan for the previously discovered set of pilot channels, rather than every legally permitted frequency.

Mobile Call Origination

When a customer initiates a call from a mobile station, this is referred to as call origination. This is typically performed after a subscriber has entered a telephone number via the number buttons and they have pressed the SEND button.

When a user initiates a call to a CDMA network, the MS first sends the dialed digits along with the phone's identification information to a nearby base station. After the dialed digits have been received and the MS has been authorized for service, the MSC will seize an outside line (trunk) and dial the indicated number. The CDMA network will then command the MS to tune to a specified radio carrier frequency and channel code for which the call will be connected. The MS will change code channels and conversation may begin.

During the conversation, the Base Transceiver Station is continually measuring the signal strength of the received radio waves from the mobile station's transmitter. In addition, all of the digital information transmitted over the radio link consists of two portions. One portion is the actual information of significance, such as the encoded speech or the call control messages that cause the MS to re-tune to another frequency or other actions. The other portion is a smaller set of data bits called an error-detection code. There is a method used at the radio receiver to examine the information bits and the error-detection code bits for consistency. When these two portions are not mathematically consistent, this detects that errors have occurred during transmission via the radio channel. In many cases, the number of erroneous data bits can be estimated with good accuracy. Then the bit error rate (BER) can be computed, which is the ratio of erroneously received bits to the total of all received bits. Due to radio channel imperfections, about 1% of the data bits transmitted are (about 1 erroneous bit out of each 100 bits received) received in error. When there is a 2, 3 or even up to 5% BER for a short interval of time, the voice codec is still able to produce sound with reasonable accuracy. But when the BER goes much above 5% for a long enough time, the sound output will be unacceptably bad, as is known from prior measurements.

Call Handoff

Call handoff is the process of transferring a call between base stations. Handoff is typically called handover in Europe. Handoff is necessary because mobile stations often moves out of range of one base station and into the radio coverage area of another base station.

The CDMA network has several advantages for handoff when compared to analog systems. Because the CDMA network is digital and divided in different codes rather than frequencies, it is possible for the mobile radio to simultaneously communicate with more than one base station at the same time. This allows the gradual transfer of a call from one cell site to another. This process of transfer is called soft handoff.

In the CDMA network, there are several additional items of information that are not available in older analog cellular systems that can be used during handoff. The control channel codes are continually broadcast from each cell site. This allows the mobile radio to quickly locate the other control channels (e.g. paging channel and access channel). During continuous communication between the MS and the base station (e.g. conversation), the MS can continuously report back its quality of its received signal. This information combined with the base station's sensing of the signal quality received from the MS helps to determine when the call should be transferred to another cell site.

The base system also knows which adjacent cells have idle radio channels available as a handover target, and which do not. The control computer in the BSC (or in the

MSC, as the case may be) selects the set of adjacent cells which have idle available channels, and from this set it selects that cell which has the best combination of signal strength and BER. A suitable channel (carrier frequency and code) is assigned in that cell as the target, and the MS is commanded to re-tune and change its code to use that channel. For a period of time, the MS communicates with the current base station and the base station where a potential handoff may occur. There is no lost information, and no gap in the speech from the voice codec. This is called a "seamless" handover.

The basic call handover process begins when a MS receives a list of radio carrier channels from the base station that is communicating with which allows it to measure the signal strength of nearby base stations. After the MS measures the quality of the other radio carrier channels, it returns this information to the serving base station. Using this information and information from neighboring base stations, the serving base station sends a message which instructs the MS to monitor a new PN channel code of an adjacent base station. The MS begins transmission on the new channel. After the MS has connected to the new base station, the MSC uses the audio signals from both base stations. When the signal quality at the new targeted base station is much higher than the original base station, the MSC disconnects the audio path from the original base station and the handoff is complete.

Ending a Call

Eventually, one of the two people involved in the telephone conversation hangs up the telephone (on the land end) or presses the END button on the mobile end. This causes an exchange of messages over the radio link which requests a disconnection message along with an acknowledgment message for an intentional disconnection. The system design is very robust in this situation. The system requires messages to be repeated to confirm a disconnect order. This helps to prevent an accidental disconnection.

After the call is disconnected, the MS starts scanning again to find the best pilot channel and be ready for another call. The base station marks the previous channel as free and ready for another use by another conversation.

Receiving a Call on a Mobile

Receiving a call on a mobile phone is called call termination. A mobile terminated call is essentially similar to the mobile originated call just described, except for the beginning steps which involve alerting the mobile station of an incoming call (called paging). The paging process begins when another caller dials the telephone number of the mobile station. This results in an inquiry to the HLR (customer database) of the home MSC switch. The HLR responds to the request with a message that includes the identification number of the MS along with an indication of the last registered location of the MS. If the mobile phone is operating in a visited system, the HLR response includes the system identification and routing information of the visited system. The system uses the mobile phone identification information to send a page message to the MS. The page message is sent to the MS on a radio carrier channel of the base station where the MS was located last.

The MS identification information is called the mobile identification number (MIN). Because the MIN is composed of many digits, systems typically use an abbreviated form of the paging message. The temporary mobile identification number (TMIN) is assigned to the mobile phone when it first registers in a system (typically during initialization).

The basic process for receiving a call on the CDMA network requires the MS to continuously monitor a paging channel until it hears its identification number. After it hears its specific paging number (page message), the MS will then request service from the CDMA network indicating in its request that it is responding to a page message. After the system validates the MS identification information, it will assign it to a radio carrier channel.

References:

1. CDMA Development Group (CDG), 31 Jan 99.
2. Strategis Group, "World Cellular and PCS Markets study, Washington DC, 1997.

Chapter 2
Analog Cellular

The IS-95 CDMA system was designed to allow dual mode operation with the advanced mobile phone system (AMPS). This chapter describes the basic analog AMPS system.

History

When the current U.S. cellular system was introduced in 1983, it was termed Advanced Mobile Phone Service (AMPS), now defined by the Electronics Industries Association (EIA) specification EIA-553, Base Station to Mobile Station Compatibility Standard. To work in the U.S. system, mobile and base station units must be manufactured to this specification.

AMPS systems operate in over 72 countries [1]. The AMPS standard continues to evolve to allow advanced features such as increased standby time, narrowband radio carriers, and anti-fraud authentication procedures.

The history of analog cellular began in 1971 when AT&T proposed a cellular radio telephony system [2] that could meet the FCC's requirement for a wireless system to serve a large number of customers (called subscribers) with a limited amount of radio spectrum. The AT&T cellular system proposal was the backbone of the U.S. commercial cellular system that started service in Chicago, October 1983. Since that time, cellular radio telephony has evolved into the Analog FM (EIA-553) cellular system we have today. This system was originally called the Advanced Mobile

Phone Service (AMPS) system. Although the name of the analog system has changed, the EIA-553 system is commonly referred to as the "AMPS" system.

To provide for competition in the United States, the FCC divided each cellular service area radio spectrum into two frequency bands which are assigned by two cellular companies, called A and B carriers. There are 734 cellular service areas in the U.S., each has an A and B carrier. The A carrier does not have a controlling interest in the local telephone company; the B carrier (often a Bell operating company) can have a controlling interest in a local telephone company.

AMPS System Overview

The AMPS radio system has dedicated control channels and voice channels. Mobile telephones use one of 21 dedicated control channels to listen for pages and compete for access. The control channels continuously send system identification information and access control information. Although the control channel data rate is 10 kbps, messages are repeated 5 times which reduces the effective channel rate to below 2 kbps. This allows a control channel to send 10 to 20 pages per second.

Signaling on the radio voice channel is performed by transferring messages on a dedicated control channel and by blank and burst signaling on a voice channel. In addition to the control messages, one of three supervisory tones (called SAT) which are approximately 6 kHz are combined with the voice information sent to all the checking for continuous radio connection.

Dual Mode Telephone

The IS-95 CDMA system offers the option of analog or dual mode operation. A mobile telephone (typically called a mobile station) contains a radio transceiver, user interface, and antenna assembly (see figure 2.1) in one physical package. The radio transceiver converts audio to a radio frequency (RF) signal and RF signals into audio. A user interface provides the display and keypad which allow the subscriber to communicate commands to the transceiver. The antenna assembly couples RF energy between the electronics within the mobile telephone and the outside "air" for transmission and reception. Figure 2.1 shows a block diagram of a dual mode mobile telephone that allows communication on either AMPS radio carriers or CDMA radio carriers.

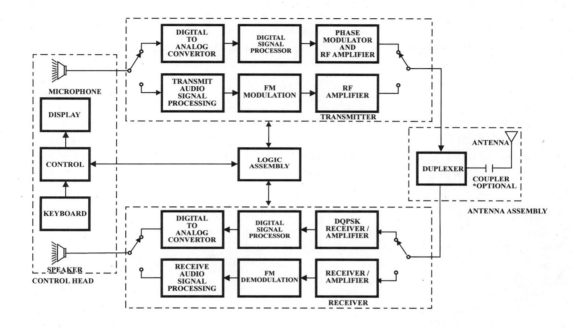

Figure 2.1, Dual Mode Mobile Telephone Block Diagram

Analog mobile cellular telephones have many industry names. These names sometimes vary by the type of cellular radio. Handheld cellular radios are often referred to as "portables". Cellular radios that are installed in cars are typically called "mobiles". Cellular radios mounted in bags are often called "bag phones". In most cases, these three types and sizes also correspond to three distinct maximum power levels: 600 milliwatts, 1.8 watts, and 3 watts, or classes I, II and III. We will refer to any type of mobile cellular radio as a mobile station (MS). The evolution of portable cellular telephones has resulted in approximately 20% reduction in weight [3], and 24% reduction in cost [4] each year over the past 5 years.

Each mobile station must have distinct signaling identification when operating in a cellular system and contain feature options specific to the customer. To make each mobile telephone unique, several types of information is stored in its internal memory. In the United States, this internal memory is called a Number Assignment Module (NAM). The NAM contains the Mobile Identification Number (MIN) which is the telephone number, home system identifier, access classification, and other customer features. The internal memory which stores the telephone number and system features can be modified either by changing a chip stored inside the mobile telephone or by programming the phone number into memory through special key-pad instructions. One of the new features of CDMA mobile radios is the ability to change the programming information by sending a message over the radio carrier channel.

Mobile telephones also contain a unique Electronic Serial Number (ESN) which is not supposed to be changed. If the ESN could be easily changed, it would be possible to duplicate (called cloning) another mobile telephone's identification to make fraudulent calls. Because duplications of mobile telephone numbers and ESN's is technically possible, advanced authentication programs which validate pre-stored information have been created to provide a more reliable unique identification system.

Initially, information stored in a NAM was programmed into a standard Programmable Read Only Memory (PROM) chip. Because of the cost of the chips and that special programming devices were required, manufacturers now make the NAM information programmable via the handset keypad. The information is stored internally in an electrically alterable PROM (EPROM). This is also referred to as a non-volatile memory, since the information contents stay intact even if power is not available, such as when a battery is replaced.

System Overview

The AMPS cellular system provides telephone service to many customers through duplex radio carriers, frequency reuse, cost effective capacity expansion, and coordinated system control. To conserve the limited amount of radio spectrum, cellular systems reuse the same channels many times within a geographic coverage area. The technique, called frequency reuse, makes it possible to expand system capacity by increasing the number of channels that are effectively available for subscribers. As the subscriber moves through the system, the Mobile Telephone Switching Office (MTSO) centrally transfers calls from one cell to another and maintains call continuity. In fact, without frequency reuse, it would not be economically feasible to provide cellular or PCS service, unless all other radio frequency bands (broadcasting, emergency radio systems, ship to shore, military, etc.) were shut off and their spectrum capacity were also used for cellular/PCS.

Frequency Allocation

In 1974, 40 MHz of spectrum was allocated for cellular service [5] which provided only 666 channels. In 1986, an additional 10 MHz of spectrum was added to facilitate expansion [6] which expanded the system to 832 channels.

AMPS radio carriers are frequency duplex with its channels separated by 45 MHz. The control channel and voice channel signaling is transferred at 10 kbps. AMPS cellular phones have three classes of maximum output power. A class 1 mobile telephone has a maximum power output of 6 dBW (3 Watts), class 2 has a maximum output power of 2 dBW (1.6 Watts), and the class 3 units are capable of supplying only -2 dBW (.6 Watts). The output power can be adjusted in 4 dB steps and has a minimum output power of -22 dBW (approximately 6 milliwatts).

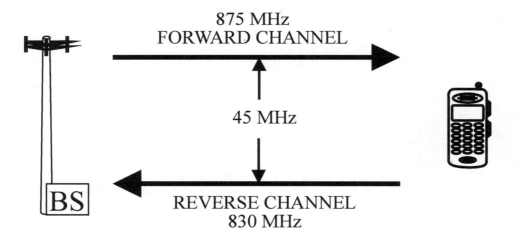

Figure 2.2, Duplex Radio Carrier Spacing

To allow simultaneous transmission and reception (no need for push to talk), the base stations transmit on one set of radio carriers, called forward channels and they receive on another set of channels, called the reverse channels. The transmit and receive channels assigned for a particular cell are separated by a fixed amount of frequency. Figure 2.2 (a) displays a base station transmitting to the mobile telephone at 875 MHz on the forward channel. The mobile telephone then transmits to the base station at 830 MHz on the reverse channel. Figure 2.2 (b) shows the base station transmitting at 890 MHz resulting in the mobile telephone transmitting at 845 MHz.

Frequency Reuse

In early mobile radio systems, one high-power transmitter with a modest allocation of frequency spectrum served a large geographic area. Because each cellular radio

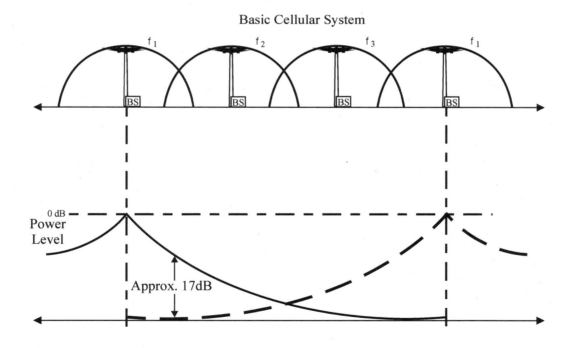

Figure 2.3, Frequency Reuse

requires a certain bandwidth, the resulting limited number of radio carriers kept the serving capacity of such systems low. The customer demand for the few available channels was very high. For example, in 1976, New York City had only 12 radio carriers to support 545 subscribers and a two-year long waiting list of typically 3,700 [7].

antenna, channel

To increase the number of radio carriers in where the frequency spectrum allocation is limited, cellular providers must reuse frequencies. One strategy for reusing frequencies relies on the fact that signal strength decreases exponentially with distance, so subscribers who are far enough apart can use the same radio carrier without interference (see figure 2.3).

To minimize interference in this way, cellular system planners position the cell sites that use the same radio carrier far away from each other. The distances between sites are initially planned by general RF propagation rules, but it is difficult to

TOP VIEW

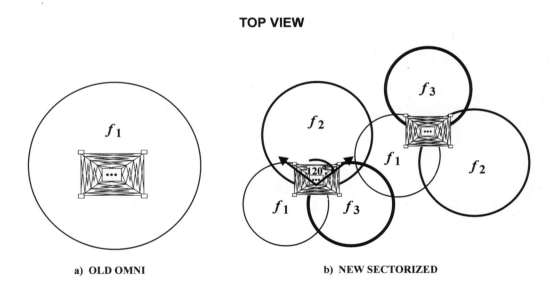

a) **OLD OMNI**   b) **NEW SECTORIZED**

Figure 2.4, Cell Site Sectorization

account for enough propagation factors to precisely position the towers, so the cell site position and power levels are usually adjusted later.

An acceptable distance between cell sites that use the same radio carrier frequency is determined by a distance to radius (D/R) ratio. The D/R ratio is the ratio of the distance (D) between cells using the same radio frequency to the radius (R) of the cells. For the AMPS system, a typical D/R ratio is 4.6: a channel used in a cell with a 1 mile radius would not interfere with the same channel being reused at a cell 4.6 miles away.

Capacity Expansion

As cellular systems mature, they must serve more subscribers, either by adding more radio carriers in a cell, or by adding new cells. To add radio carriers, cellular

By using sectorize ant.

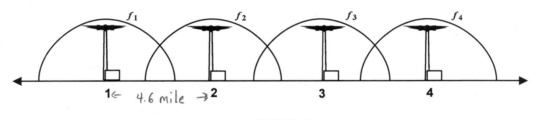

a) ORGINIAL

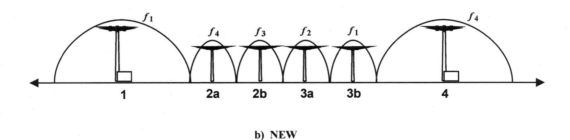

b) NEW

Figure 2.5, Cell Splitting

systems use several techniques in addition to strategically locating cell sites that use the same frequencies. Directional antennas and underlay/overlay transmit patterns improve signal quality by focusing radio signals into one area and reducing the interference to other areas. The reduced interference allows more frequency reuse. Directional antennas can be used to sector a cell in to wedges so that only a portion of the cell area (e.g. 1/3 or 120 degrees) is used for a single radio carrier. Such sectoring reduces interference with the other cells in the area. Figure 2.4 shows cells that are sectored into three 120 degree sectors.

Another technique, called cell splitting, helps to expand capacity gradually. Cells are split by adjusting the power level and/or using reduced antenna height to cover a reduced area (see figure 2.5). Reducing a coverage area by changing the RF boundaries of a cell site has the same effect as placing cells farther apart, and allows new cell sites to be added. However, the boundaries of a cell site vary with the terrain and land conditions, especially with seasonal variations in foliage. Coverage areas actually increase in fall and winter as the leaves fall from the trees.

Current analog systems serve only one subscriber at a time on a radio carrier. The number of radio carriers determines the maximum system capacity. Because a typical subscriber uses the system for only a few minutes each day, many subscribers share a single channel. Typically, 20 - 32 subscribers share each radio carrier [8], depending upon the average talk time per hour per subscriber. Generally, a cell with 50 radio carriers can support 1000 - 1600 subscribers.

When a cellular system is first established, it can effectively serve only a limited number of callers. When that limit is exceeded, callers experience too many system busy signals (known as blocking) and their calls cannot be completed. More callers can be served by adding more cells with smaller coverage areas - that is, by cell splitting. The increased number of smaller cells provides more available radio carriers in a given area because it allows radio carriers to be reused at closer geographical distances.

System planning must also account for present and future coverage requirements. After the cellular service provider is granted a license, the cellular service providers typically have only a few years to provide coverage to almost all of their licensed territory [9]. To accomplish this, and to ensure that the system will be efficient and competitive, cellular carriers must plan and design the system in advance.

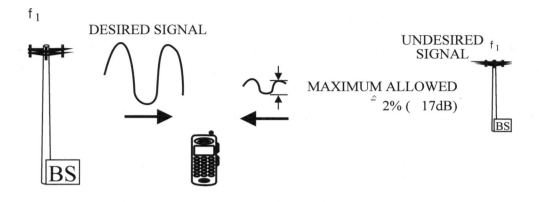

Figure 2.6, Co-Channel Interference

Radio Interference

Radio interference limits the number of radio carriers that can be used in a single cell site and how close nearby cell sites that use the same frequency can be located together. The main types of interference are co-channel, adjacent channel, and alternate channel interference.

Co-channel Interference

Co-channel interference occurs when two nearby cellular radios operating on the same radio carrier interfere with each other. Co-channel interference at a particular location can be measured by comparing the received radio signal power (signal strength) from the desired signal with the signal strength of the interfering signal. Today's analog systems are designed to assure that interfering signal strength remains approximately less than 2 percent of desired signal strength. At this level, the desired signal is nearly undistorted.

To minimize interference, cellular carriers frequently monitor the received signal strength by regularly driving test equipment throughout the system. This testing determines whether the combined interference from cells using the same channel exceeds the 2 percent level of the desired channel (which is 17 dB below the desired signal). This information is used to determine whether radio carriers and/or power levels at each cell need to be changed. Figure 2.6 shows how co-channel interference occurs.

Radio technology that provides a higher tolerance to co-channel interference (i.e. exceeding 2 percent with no distortion) would allow system operators to reuse the same frequencies more often, thus increasing the system capacity. This higher tolerance is an advantage of next-generation digital cellular technologies.

Adjacent Channel Interference

Adjacent channel interference occurs when one radio carrier interferes with a channel next to it (e.g. radio carrier 412 interferes with radio carrier 413). Each radio carrier has a limited amount of bandwidth (10 kHz to 30 kHz wide), but some radio

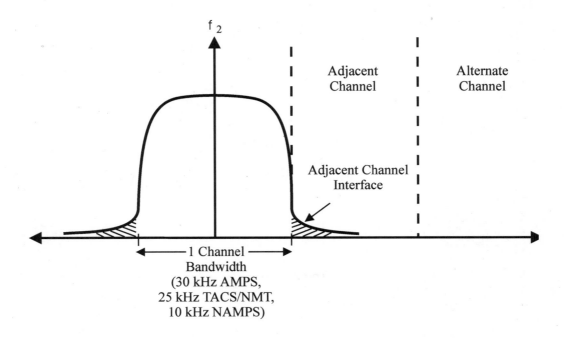

Figure 2.7, Adjacent Channel Interference

energy is transmitted at low levels outside this band. A cellular radio operating at full power can produce enough low-level radio energy outside the channel bandwidth to interfere with cellular radios operating on adjacent channels. Because of alternate channel interference, radio carriers cannot be spaced adjacent to each other in a single cell site (e.g. channel 115 and 116). A channel separation of 3 channels is typically sufficient to protect most radio carriers from adjacent channel interference. However, for frequency planning reasons, the radio carrier frequencies at each cell site are selected so they are typically separated by 21 channels from other radio carriers in that base station or sector. Figure 2.7 displays adjacent channel interference.

Alternate Channel Interference

Alternate channel interference occurs when radio energy from a transmitted signal occurs outside its designated frequency band which is located two radio carrier bandwidths (e.g. radio carrier 412 interferes with radio carrier 414) away interferes with the desired signal.

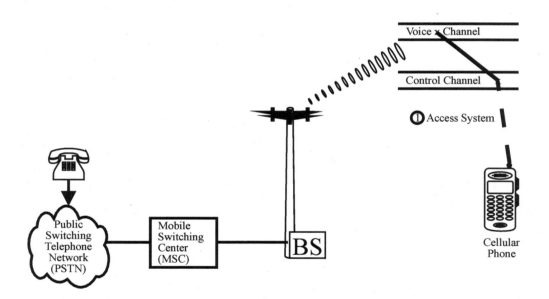

Figure 2.8, Control Channels and Voice Channels

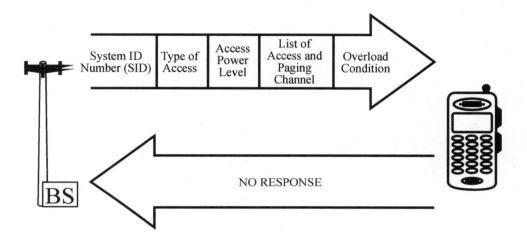

Figure 2.9, Cellular System Broadcast Information

Basic Cellular Operation

In early mobile radio systems, a mobile telephone scanned the limited number of available channels until it found an unused radio carrier. Once the unoccupied radio carrier was found (typically by a pilot tone), the mobile telephone was allowed it to initiate a call. Because the analog cellular systems in use today have hundreds of radio carriers, a mobile telephone cannot scan them all in a reasonable amount of time. To quickly direct a mobile telephone to an available channel, some of the available radio carriers are dedicated as control channels. Most cellular systems use two types of radio carriers: control channels and voice channels. Control channels only carry digital messages and signals that allow the mobile telephone to retrieve system control information and compete for access to the system. Control channels never carry voice (in the AMPS system). Voice channels are primarily used to transfer voice information, but also send and receive some control messages. Figure 2.8 displays that each cell site has at least one radio carrier designated as a control channel to coordinate access to its radio carriers that are designated as voice channels. After the access to a voice channel has been authorized, the control channel sends out a channel assignment message that commands the mobile telephone to tune to a voice channel.

When a mobile telephone is first powered on, it initializes itself by scanning the predetermined set of control channels and then tuning to the strongest radio carrier. Figure 2.9 shows that during this initialization mode, it retrieves system identification and setup information.

After initialization, the mobile telephone enters the idle mode and waits to be paged for an incoming call and senses if the user has initiated (dialed) a call (requested system access). When a call begins to be received or initiated, the mobile telephone enters system access mode to try to access the system via a control channel. When it gains access, the control channel sends an initial voice channel designation message indicating an open voice channel. The mobile telephone then tunes to the designated voice channel and enters the conversation mode. As the mobile telephone operates on a voice channel, the system uses Frequency Modulation (FM) similar to commercial broadcast FM radio. To send control messages on the voice channel, the voice information is either replaced by a short burst (blank and burst) message or in some systems, control messages can be sent along with the audio signal.

Access

A mobile telephone's attempt to obtain service from a cellular system is referred to as "access". Mobile telephones compete for access by sending service request messages on a control channel. Access is attempted when a command is received by the mobile telephone indicating the system needs to service that mobile telephone (such as a paging message indicating a call to be received) or as a result of a request from the user to place a call. The mobile telephone gains access by monitoring the busy/idle status of the control channel both before and during transmission of the access attempt message. If a control channel is not busy, the mobile telephone begins to transmit a service request. Simultaneously, the base station monitors the channel's busy status. Transmissions must begin within a prescribed time limit after the mobile station finds that the control channel access is free, or the access attempt is stopped on the assumption that another mobile telephone has possibly gained attention of the base station control channel receiver. Figure 2.10 shows a sample access process.

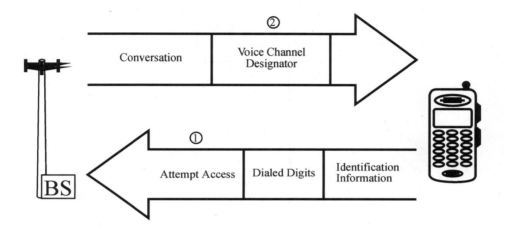

Figure 2.10, Cellular System Call Origination Radio carrier Access

If the access attempt is successful, the system sends out a channel assignment message commanding the mobile telephone to tune to a cellular radio carrier voice channel. Figure 2.11 displays the access process when a call is placed from the mobile telephone to the cellular system (called "origination"). The access attempt message is called a Call Setup message and it contains the dialed digits and other information. If a voice channel is available, the system will assign a voice channel by sending a voice channel designator message. If the access attempt fails, the mobile telephone waits a random amount of time before attempting access again. The design of the system minimizes the chance of repeated collisions between different mobile stations which are both trying to access the control channel, since each one waits a different random time interval before trying again if they have already collided on their first, simultaneous attempt.

An access overload class (ACCOLC) code is stored in the mobile telephone's memory which can inhibit it from transmitting when the system is busy serving many subscribers. When the system is very busy, an access overload class category is sent on the control channel. If the ACCOLC matches the stored access overload class in the mobile telephone, it is inhibited from attempting to access the cellular system. This process allows the cellular system to selectively reduce the number of access

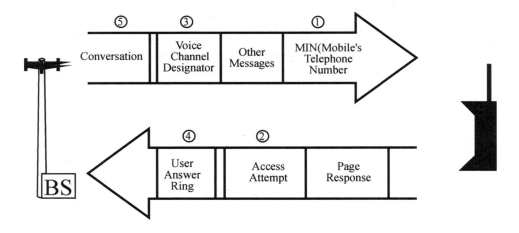

Figure 2.11, Cellular System Paging

attempts and only allow particular groups of mobile telephones to access the system. Different ACCOLC are assigned to specific types of mobile stations. Specific (higher level) access classes are assigned to mobile telephones that are used by emergency personnel.

√ Paging

To receive calls, a mobile telephone is notified of an incoming call by a process called paging. A page is a control channel message which contains the telephone's Mobile Identification Number (MIN or telephone number of the desired mobile phone) and it responds automatically with a system access message of a type called a Page Response. This indicates that an incoming call is to be received. After the mobile telephone receives its' own telephone number, the mobile telephone begins to ring. When the customer answers the call (user presses SEND), the mobile telephone transmits a service request to the system to answer the call. It does this by sending the telephone number and an electronic serial number to provide the users identity. Figure 2.11 shows that if the mobile telephone is paged in the system and wishes to receive the call (user presses SEND), it responds to the page by attempting access to the system.

Conversation

After a mobile telephone has been commanded to tune to a radio carrier voice channel, it sends mostly voice information. Periodically, control messages may be sent between the base station and the mobile telephone. Control messages may command the mobile telephone to adjust its' power level, change frequencies, or request a special service (such as three way calling).

Discontinuous Transmission

To conserve battery life, a mobile phone may be permitted by the base station to only transmit when it senses the mobile telephone's user is talking. When there is silence, the mobile telephone may stop transmitting for brief periods of time (several seconds). When the mobile telephone user begins to talk again, the transmitter is turned on again.

Handoff

Handoff is a process where the cellular system automatically switches channels to maintain voice transmission when a mobile telephone moves from one cell radio coverage area to another. The MSC's switching equipment transfers calls from cell to cell and connects the call to other mobile telephones or the public telephone network. The MSC creates and interprets the necessary command signals to control mobile telephones via base stations. This allows the switching from channel to channel as the mobile telephone moves from one coverage area to another.

Figure 2.12 shows the cellular handoff process. Initially, base station #1 is communicating with the mobile telephone (t1). Because the signal strength of the mobile telephone has decreased, it has become necessary to transfer the call to a neighboring cell, base station #2. This is accomplished by base station #1 sending a handoff command to the mobile telephone (t2). The mobile telephone tunes to the new radio carrier (428) and begins to transmit a control tone that indicates it is operating on the channel (t3). The system senses that the mobile telephone is ready to communicate on channel 428 and the MSC switches the call to base station #2 (t4). The conversation can then continue (t5). This entire process is usually accomplished in less than 1/4 of a second.

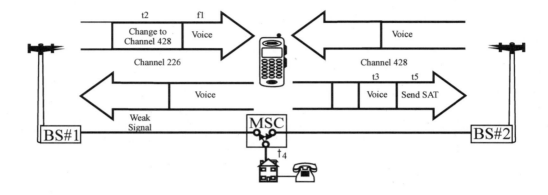

Figure 2.12, Cellular System Handoff

When a mobile telephone moves far away from the base station that is serving it, the cellular system must transfer service to a closer base station. Figure 2.13 illustrates the process. To determine when handoff is necessary, the serving base station continuously monitors the signal strength of the mobile telephone. When the mobile telephones' radio signal strength falls below a minimum level, the serving base station requests adjacent base stations to measure that radio's signal strength (step 1). The adjacent base stations adjust the frequency of their scanning receiver to monitor the signal strength of the mobile telephones' current operating channel. When a closer adjacent base station measures sufficient signal strength (step 2), the serving base station commands the cellular radio to switch to the new base station (step 3). After the mobile telephone starts communicating with the new base station, the communication link carrying the landline voice path is switched to the new serving base station to complete the handoff (step 4).

√ RF Power Control

Mobile telephones are typically classified by their maximum amount of power output, called the "power class". Mobile telephone power output is adjusted by commands received from the base station to reduce the transmitted power from the

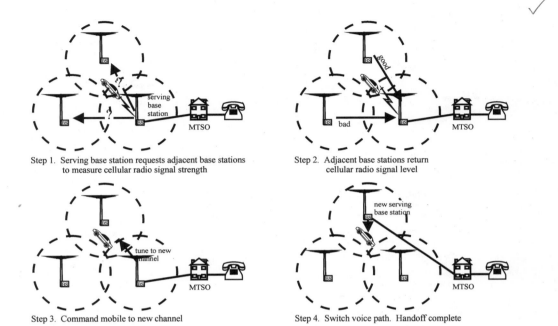

Step 1. Serving base station requests adjacent base stations to measure cellular radio signal strength

Step 2. Adjacent base stations return cellular radio signal level

Step 3. Command mobile to new channel

Step 4. Switch voice path. Handoff complete

Figure 2.13, Handoff Messaging

mobile telephone in smaller cells. This reduces interference to nearby cell sites. As the mobile telephone moves closer to the cell site, less power is required from the mobile telephone and it is commanded to reduce its transmitter output power level. The base station transmitter power level can also be reduced although the base station RF output power is not typically reduced. While the maximum output power varies for different classes of mobile telephones, typically they have the same minimum power level. Figure 2.14 shows the RF power control process.

Roaming

A home system identifier code is stored in the mobile telephone's memory. This allows the mobile telephone to compare the home system identifier code to the system identifier code that is transmitted on the serving control channel. If they do not match, it means the subscriber is operating in a visited system and the mobile tele-

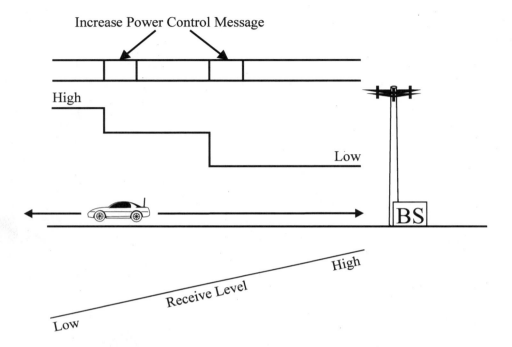

Figure 2.14, RF Power Control

phone will provide a ROAM indicator. The ROAM indication allows the subscriber to determine that they are operating outside their home area and new billing rates may apply.

Signaling

Signaling is the transferring of control messages between two points. There are two basic parts of the signaling process: the physical transport of the message and the actual content of the message. Control messages are sent on radio control channels, radio voice channels and between the network parts of the cellular and telephone system.

Radio Control Channels

Most cellular systems have dedicated control channels that carry several types of messages to allow the mobile telephone to listen for pages and compete for access. These messages include:

<u>Overhead messages</u> which continuously communicate the system identification (SID) number, power levels for initial transmissions, and other important system registration information
<u>Pages</u> that identify to a particular mobile telephone that a call is to be received
<u>Access information</u> which is the information exchanged between the mobile telephone and the system to request service
<u>Channel assignment commands</u> that establish the radio carriers for voice communications.

The control channel sends information by Frequency Shift Keying (FSK). To allow self-synchronization, the information is Manchester encoded which forces a frequency shift (bit transition) for each transmitted bit of information [10]. Control messages (instruction orders) are sent as word messages and a control messages may be composed of one or more words.

To help coordinate multiple mobile telephones accessing the system, busy idle indicator bits are typically interlaced with the other message bits. Before a mobile telephone attempts access to the system, it checks the busy/idle bits to see if the control channel is serving another mobile telephone. This system is called Carrier Sense Multiple Access (CSMA) and it helps to avoid collisions during access attempts.

When a mobile telephone begins to listen to a control channel, it must find the beginning of messages so they can be decoded. Messages are preceded with an alternating pattern called a dotting sequence which is easy to sense and identifies a message will follow. Following the dotting sequence, a unique sequence of bits (call a synchronization word) are sent that allows the mobile telephone to match the exact start time of the message.

Radio carriers can have rapid signal level fades which introduce errors, so the message words are repeated several times to ensure reliability. Of the repeated words, the mobile telephone can use a majority vote system to eliminate corrupted messages. Message and signaling formats on the control channels vary between forward and reverse channels. The forward channel is synchronous and the reverse channel is asynchronous.

Forward Control Channel

On the forward control channel, several message words follow a dotting and synchronization word sequence. Each word has error correction/detection bits that are included so the data content can be verified and possibly corrected if received in error.

Reverse Control Channel

On the reverse control channel, words follow a dotting and synchronization word sequence. Because the reverse channel is randomly accessed by mobile telephones, the dotting sequence in the reverse direction is typically longer than the dotting sequence in the forward direction. Each reverse channel word has error detection and correction bits. Messages are sent on the reverse channel in random order and typically coordinated using the Busy/Idle status from a forward control channel.

Radio Carrier Voice Channels

After a mobile telephone is assigned a radio carrier voice channel, voice and control information must share the same radio carrier. Brief control messages that are sent on the voice channel include:

Handoff messages that instruct the mobile telephone to tune to a new channel
Alert message tells the mobile telephone to ring when a call is to be received
Maintenance command messages monitor the status of the mobile telephone
Flash requests a special service from the system (such as 3 way calling)

The radio carrier voice channel typically transfers voice information between the mobile telephone and the base station. Signaling information must also be sent to allow base station control of the mobile telephone. Signaling on the voice channel can be divided into in band and out of band signaling. In-band signaling occurs when audio signals between 300-3000 Hz either replace or occur simultaneously with voice information. Out-of-band signals are above or below the 300-3000 Hz range, and may be transferred without altering voice information. Signals sent on the voice channel include a pilot or Supervisory Audio Tone (SAT), Signaling Tone (ST), Dual Tone Multi-Frequency (DTMF), and blank and burst FSK digital messages.

The supervisory tone provides a reliable transmission path between the mobile telephone and base station, and is transmitted along with the voice to indicate a closed

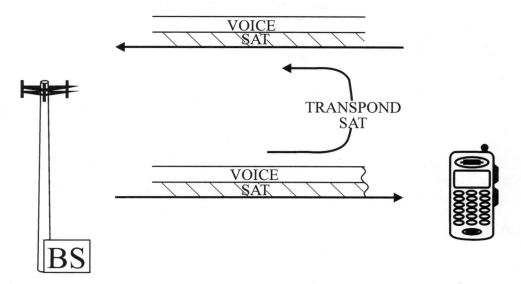

Figure 2.15, Transponding SAT

loop. The tone functions are much like the current/voltage functions used in land line telephone systems to indicate that a phone is off the hook [11]. The supervisory tone may be one of the several frequencies (around 6 kHz) and this tone is different for nearby cell sites. If the supervisory tone is interrupted for longer than about 5 seconds, the call is terminated. Figure 2.15 shows how a supervisory tone is transponded (repeated back) to the base station.

The loss of SAT signal indicates that the radio connection has been interrupted or interference has occurred. The use of different supervisory tone frequencies in adjacent base stations is also used to ensure that radio carriers from nearby base stations that are operating on the same frequency do not control the mobile telephone. When the sensing of a transponded SAT tone has been lost (possibly by a signal fade) or an incorrect SAT frequency has been received (from an interfering signal), the mobile telephones' audio signal is temporarily muted.

Re-transmission of the supervisory tone can also be used to locate the mobile telephone's position. An approximate propagation time can be calculated by comparing the phase relationship between the transmitted and received supervisory tones. This propagation time is correlated to the distance from the base station. However, multipath propagation (radio signal reflections) makes this location feature inaccurate and only marginally useful [12]. Only the re-transmission of the supervisory tone as a pilot tone is critical to operation.

A signaling tone (ST) is used in some systems to indicate a call status change. It confirms messages sent from the base station and is similar to a public telephone status change of going on or off hook [13].

Touch-tone (registered trademark of AT&T) signals may be sent over the voice channel. DTMF signals are used to retrieve answering machine messages, direct automated PBX systems to extensions, and a variety of other control functions. The voice channel can transmit DTMF tones, but varying channel conditions can alter the expected results. In poor radio conditions and a fading environment, the radio path may be briefly interrupted, sometimes sending a multiple of digits when a key was depressed only once.

Blank and Burst Messages

When signaling data is about to be sent on the voice channel, audio FM signals are inhibited and replaced with digital messages. This interruption of the voice signal is normally so short (less than/second) that it is often unnoticed by the mobile telephone user. Like control channel messages, these messages are typically repeated a multiple of times and a majority vote is taken to see which messages will be used.

To inform the receiver that a digital signaling message is coming, a bit dotting

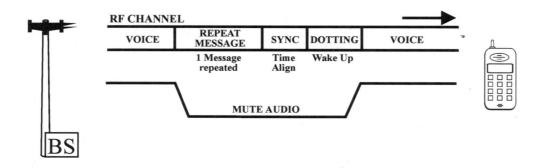

Figure 2.16, In Band Voice Channel Message

sequence is sent preceding the message. After the dotting sequence gets the attention of the receiver, a synchronization word follows which identifies the exact start of the message. Figure 2.16 shows how a voice channel message is sent.

Blank and burst signaling differs on the forward and reverse voice channels. On the forward voice channel, messages are repeated more times to ensure control information is reliable even in poor radio conditions. It is likely that messages will be sent in poor radio conditions as handoffs messages often occur when the signal is very weak.

Network

Signaling commands must be passed between base stations, MCS, and the public telephone network. These commands are for the maintenance, control, and administration of the network.

Signaling between the base station and MSC is performed on one of the multiple channels connected between base stations and the switching center. This control channel is designated exclusively as a control (data) channel. The MSC uses this data link to send commands to the base station and receive information about the calls in progress. The MSC uses this capability to switch calls between cell sites (handoff) and the public telephone network. Base stations also use software that is sometimes changed when new features or test software becomes available. The data link is used to download this new software.

Early cellular systems were connected to the public switched telephone network in a way that was similar to a standard telephone line. Unfortunately, this type of connection does not provide much information about the call status. Most modern cellular systems connected to the public telephone network in a similar way as the standard public (landline) telephone switch connects to other telephone switches. This allows a cellular system to receive and send control messages that contain detailed information about the calls in process.

Reference:

1. Cellular and PCS/PCN Telephones and Systems, pp. 377-381, APDG Publishing, Fuquay, NC, 1996.
2. Dr. George Calhoun, "Digital Cellular Radio", p.50-51, Artech House, MA. 1988.
3. "The Retail Market of Cellular Telephones", 1984-1996, Herschel Shosteck Associates, Wheaton, MD.
4. Ibid.
5. William Lee, "Mobile Cellular Telecommunications Systems", p.5, McGraw Hill, 1989.
6. Ibid, p.265.
7. Ibid, p.2.
8. Personal interview, industry expert, 3 April 1999.
9. FCC Regulations, Part 22, Subpart K, "Domestic Public Cellular Radio Telecommunications Service," 22.903, (June 1981).
10. The Bell System Technical Journal, January 1979, Vol. 58, No. 1, American Telephone and Telegraph Company, Murray Hill, New Jersey.
11. Ibid, p.47.
12. Personal Interview, Ron Bohaychuk, Ericsson Radio Systems, 7 October 1990.

Chapter 3
CDMA Technology

Code division multiple access is a form of spread spectrum communications. Spread spectrum communications allows multiple users to share the same frequency band by spreading the information signal (audio or data) for each user over a wide frequency bandwidth. The spreading is regulated by either frequency hopping or by code spreading.

The IS-95 CDMA system defines a new type of digital radio channel through the use of code spreading. Each CDMA digital radio channel uses efficient modulation and coding algorithms to allow up to 64 different communications channels for each radio carrier signal.

Although the CDMA system uses one type of digital radio carrier, there are several types of several types of CDMA (coded) channels. These include a reference channel identifier (pilot), timing reference (synchronization), an alerting channel (paging), channel assignment coordination (access) and channels that transfer user data such as voice (traffic channels)

Key system attributes include increased frequency reuse, efficient variable rate speech compression, enhanced RF power control, lower average transmit power, ability to simultaneously receive and combine several signals to increase service reliability, seamless handoff, extended battery life (power saving) and advanced features.

Codes in CDMA

The spread spectrum encoding system used by IS-95 CDMA systems use orthogonal codes, also referred to as a Walsh Code (WC). Orthogonal coding is a system of spreading codes that have no relationship to each other. The system also combines these orthogonal codes with two pseudorandom noise (PN) sequences for each communication channel. There are different codes used for different types of channels. The overhead channels (control channels) have designated codes to be used while traffic channels codes are selected at the time the transmission or when a call is originated.

Orthogonal Codes

← 64 dif communicate channel for ea radio carrier signal

There are 64 WC, each having a of length 64 bits used in CDMA systems. On the forward link, walsh codes are used to separate the channels. The reverse link channel generation uses the WC for orthogonal modulation, also known as orthogonal signaling. Orthogonal describes the property of the code where the addition of each does not create interference to other codes. However, in multi-path environment, perfect orthogonality is difficult to achieve.

Pseudorandom Noise (PN) Codes

Each IS-95 CDMA physical radio channel can be divided into 64 separate logical (WC coded) channels. A few of these channels are used for control, and the remaining channels carry voice information and data. Because CDMA transmits digital information combined with unique codes, each logical channel can transfer data at different rates (e.g. 4800 b/s, 9600 b/s).

The CDMA system uses two types of PN code sequences; long and short sequences. The PN codes help the MS to time synchronize with the BS and to uniquely identify the MS and BS channels. The process of time shifting the PN codes is referred to as masking.

The short code is 32,768 bit long and is used for quadrature spreading on both the forward and reverse links. On the forward link, the short code is masked to identify the cell or sector in addition to the quadrature spreading. This is masking of the short code is referred to as a pilot offset or PN offset.

The long code sequence is 4400 billion bits long and is used for separating reverse link channels and data scrambling on the forward link. The long code is masked by the Electronic Serial Number (ESN) of the MS or a unique cell address.

Channel Coding and Modulation

There are different types of channel coding and modulation on the forward and reverse channels. Channel coding involves speech data compression, error protection coding and adding channel codes.

Forward Link Channel Coding

Figure 3.1 shows a basic block diagram of a forward link (base station to mobile) voice channel modulation process. This diagram shows that an audio signal is digitized to 64 kbps. This is supplied to the voice coder (Vocoder). Error protection bits and repetition bits (discussed later) are then added. These bits are then interleaved (alternated in time) to avoid the effects of group errors due to radio signal distortion. The bits are then randomized by the PN code, primarily for voice privacy. The error protection coded data signal is then spread (multipled) by the orthogonal codes to create a high speed data signal of 1.228 million information bits per second. This information signal is spread again by the long PN code. Finally, the data is sent to the modulator where it modulates the RF carrier for radio transmission.

Voice Encoder/Decoder (Vocoder): Compresses the voice data samples into a 20-msec variable rate frames. The data rate is based on speech activity. The data from the vocoder is referred to as bits.

Forward Error Correction (FEC) and Repetition: Builds redundancy into the data signal to reduce the errors at the receiver. The data from the FEC block is referred to as symbols and the data rate is now 19.2 ksp/s (kilo-symbols per second). The symbols are repeated to maintain a constant symbol rate through the remainder of the channel modulation. The amount of repetition depends on the vocoder at lower energy output.

Interleaving: Scrambles the 20 msec frame of data in a predetermine manner preventing the loss of consecutive data symbols.

Scrambling: Applies a pseudorandom noise (PN) code randomizing the data and providing for voice privacy.

Spreading: Spreads each symbol with a 64 digit orthogonal WC code. This code separates the data from other signals in the same cell. The spreading increases the data rate by a factor of 64. The data from the spreading block is referred to as chips. The data rate is now 1.2288 mcps (mega-chips per second).

Quadrature Spreading: Applies a pseudorandom noise (PN) code making the signal noise like and identifying the signal from each cell. This spreading process does not increase the data rate.

Forward Link Modulation: The CDMA system uses efficient quadrature phase shift modulation. The use of quadrature phase shift modulation allows each transmitted symbol (phase shift) to represent 2 bits of information. The base station modulator

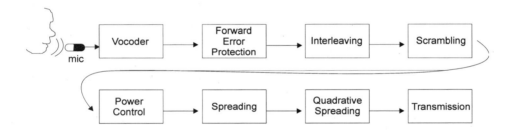

Figure 3.1, CDMA Forward Link Voice Channel Coding and Modulation

combines all the different coded digital signals into one radio channel carrier. There may be several radio channel carriers per cell site or sector within a cell site. Up to 64 coded channels can be transmitted in a single 1.23 MHz wideband signal.

RF Transmission: The modulated RF signal is supplied to a linear RF amplifier. The amount of amplification a mobile radio provides is determined by the received signal level and commands received from the base station. The RF power levels of each communication channel within a BS transmitted signal is usually equally set.

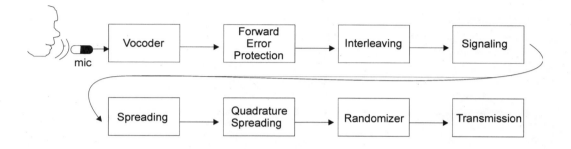

Figure 3.2, CDMA Reverse Link Voice Channel Coding and Modulatio

Reverse Link Channel Coding

Figure 3.2 shows a basic block diagram of a reverse link (mobile radio to base) voice channel coding and modulation. This diagram shows that an audio signal is digitized to 64 kbps. This is supplied to the voice coder (Vocoder). Error protection bits are then added. These bits are then interleaved (alternated in time) to avoid the effects of group errors due to radio signal distortion. The bits are then randomized by the PN code, primarily for voice privacy. The error protection coded data signal is then spread (multipled) by the orthogonal codes to create a high speed data signal of 1.228 million information bits per second. This information signal is spread again by the long PN code. A burst randomizer is used to repeat specific groups of data (discussed later). Finally, the data is sent to the modulator where it modulates the RF carrier for radio transmission.

Voice Encoder/Decoder (Vocoder): Compresses the voice data samples into a 20-msec variable rate frames. The data rate is based on speech activity. The data from the vocoder is referred to as bits.

Forward Error Correction (FEC) and Repetition: Builds redundancy into the data signal to reduce the errors at the receiver. The data from the FEC block is referred to as symbols and the data rate is now 28.8 ksps (thousand symbols per second). The symbols are repeated to maintain a constant symbol rate through the remainder of the modulation. The amount of repetition depends on the output of the vocoder.

Interleaving: Scrambles the 20 msec of data in a predetermine manner preventing the loss of consecutive data symbols.

Orthogonal Signaling: Selects one of 64 orthogonal codes to be transmitted in place of six symbols of user data. The orthogonal signaling block increases the data rate by a factor of about eleven. The data from the orthogonal signaling block is referred to as a modulation symbol (Walsh chip) and the data rate is 307.2 kcps.

Spreading: Spreads each digit of user data with four digits of a pseudorandom noise (PN) code. This code separates the signal from other signals in the reverse link. The spreading increases the data rate by a factor of four. The data from the spreading block is referred to as chips.

Quadrature Spreading: Applies a pseudorandom noise (PN) code making the signal noise like. This spreading process does not increase the data rate.

Burst Randomizer: Randomly selects which group of the repeated symbols will be transmitted.

Reverse Link Modulation: The CDMA system uses a different type of modulation for the reverse link compared to the forward link. The reverse channel uses offset quadrature phase shift keying (O-QPSK). O-QPSK differs from QPSK in that it does not require the transmitter to pass the signal through the 0 signal level when both the I and Q signals are at zero levels. This allows the mobile telephone's RF amplifier to operate more efficiently with O-QPSK because it does not need to amplify the signal as linearly (precisely) as it must with QPSK.

RF Transmission: Converts the digital signal into a RF signal. Transmission is a 1.25 msec burst. The burst transmission is a 1.23 MHz wideband signal.

CDMA Coded Channels

There are four types of coded channels originating from the cell site; pilot, sync, paging, and traffic. There are two types of coded channels originating from the MS, these are access and traffic. Figure 3.3 illustrates the code channels and the associated links. The pilot, sync, paging, and access channels carry the necessary control data while the traffic channels carry digital voice and same control data.

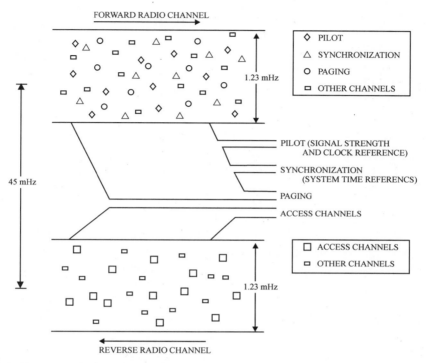

Figure 3.3, Types of CDMA Coded Channels

Pilot Channel (PC)

The Pilot Channel (PC) provides the MS with a beacon, timing and phase reference (for coherent detection), and signal strength for power control. The Pilot Code Channel (PC) is the strongest signal transmitted from the cell and always uses WC zero. The MS has fifteen seconds to acquire the PC. The PC contains no message only the PN short code, which is time shifted for cell identification.

Sync Channel (SC)

The Sync Channel (SC) provides the MS with critical time synchronization data. The message on the SC contains information necessary for the MS to align its timing. There is also information about the network air interface revision, system data, and Paging Channel data rate.

MS uses the Sync Channel (SC) for time alignment. Once the MS's timing is aligned, it will not reuse the SC until after completion of a call or it powers on again. The SC message is broken into frames and transmitted at 1200 bps. The frame is the length of the PN short code, and is time aligned with the start of the pilot PN sequence.

The SC sends WC 32 and the MS has one second to acquire it. Frame alignment with the cell PN allows a MS that has acquired the system to easily receive the SC.

Paging Channel (PgC)

The Paging Channel (PgC) contains messages with parameters that the MS needs for access and paging. The messages convey system parameters, access parameters, neighbor list, mobile directed paging messages, mobile directed orders, and channel assignment information to the MS. This channel is used to communicate with the MS when there is no call in progress.

The PgC is divided into 80 msec slots and grouped into 2048 slots. Each slot is further divided into eight half frames used to send a MS directed message, such as pages or service messages. It can be configured for either 4800 or 9600 bps data rate. Each cell may be configured to have from one to seven PgC. WC one is the default, while two through seven are optional. The MS monitors only one PgC.

Forward traffic Channel (FTC)

The Forward Traffic Channel (FTC) is a variable rate channel capable of carrying voice data, control data, or voice and control data together. The BS multiplexes MS directed messages into the FTC frames.

FTC frames are variable rate frames with the maximum transmit rates of either 9600 bps or 1.4 kbps. The FTC frames are 20 msec long and contain varying rates of data. When a FTC is setup the assigned WC will be sent to the MS using the PgC. The WC is designated for a specific MS for the duration of the call in that cell. The WC will be used in adjacent cells but the PN offset of the short code will prevent the signals from interfering.

Access Channel (AC)

The Access Channel (AC) is used to carry MS responses to commands from the BS and call origination requests. The MS communicates with the BS when there is no Reverse Traffic Channel (RTC) by using an Access Channel (AC).

There can be 32 AC for each PgC, and each AC uses a unique time shift of the PN long code. Each MS is assigned to a PgC but the AC will be randomly selected each time an access attempt is made. AC messages are sent at 4800 bps using a 20-msec frame containing 96 information bits.

Reverse traffic Channel (RTC)

Like the FTC the Reverse Traffic Channel (RTC) is a variable rate channel capable of carrying voice data, control data, or voice and control data together. The BS multiplexes MS directed messages into the FTC frames.

RTC frames are variable rate frames with the maximum transmit rates of either 9600 bps or 1.4 kbps. The RTC frames are 20 msec long and contain varying rates of data.

When a RTC is setup the public mask and the ESN, of the MS, are used to time shift the PN long code. The MS will use the PN offset for the duration of the call. Private masks can be used in place of the public mask providing an additional measure of privacy and security.

System Attributes and Features

The CDMA system has several key attributes that differentiate it from other multiple accessing schemes. Among these attributes are the, a wideband signal to mitigate the affects of fading, variable rate vocoders to reduce the level of interference, dynamic power control to provide the correct amount of power, rake receivers, soft handoff, dim-and-burst signaling, and a paging channel sleep mode.

In addition to these unique features built into the modulation scheme there is the ability to employ a frequency reuse of n=1, have a higher capacity than other technologies.

Frequency Reuse

Frequency reuse is the ability to reuse the same radio channel frequency at other cell sites within a cellular system. Because the CDMA system uses of orthogonal codes and a PN short code that have limited interference with other coded channels, it is possible for adjacent cells to use the same CDMA radio channel frequency. Figure 3.4 illustrates how the same frequency (f1) is being reused in the adjacent cells, this is known in the cellular and PCS industry as a frequency plan with n=1. In the shaded area where interference is significant, chip collisions from adjacent cells and other subscribers are more frequent, but this only reduces the number of users that can share the radio channel.

Reusing the same frequency in every cell eliminates the need for frequency planning in a CDMA system. However, pilot PN offset planning must be done in place of the frequency planning. Pilot PN offsets ensure that the received signal from one cell does not correlate with the signal from a nearby cell.

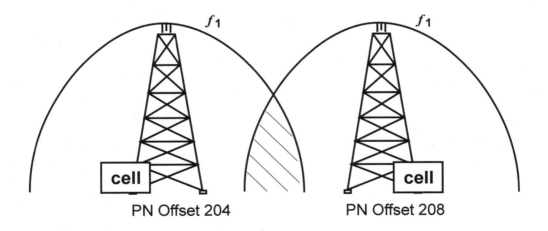

PN Offset 204   PN Offset 208

Figure 3.4, CDMA Frequency Reuse

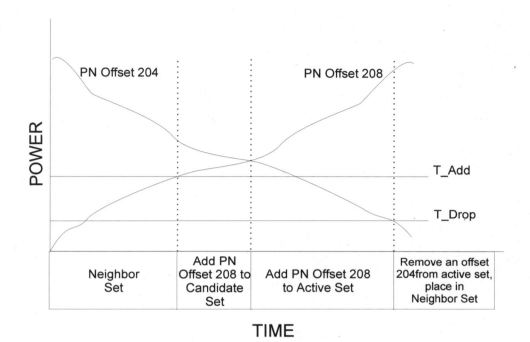

Figure 3.5, PN Offset Planning

The CDMA standard specifies 512 PN offsets, which are 64 chips apart in time. That delay equates to approximately 10 miles of signal coverage. A PN offset of 3 corresponds to 3 x 64 (chips). PN offset planning is quickly and easily accomplished using a CDMA network-planning tool. Figure 3.5 illustrates the PN offset plan for cells.

Variable Rate Voice Coder (Vocoder)

The CDMA system uses a voice coder (vocoder) that compresses the digitized voice. The amount of compression varies based on the speech activity. Speech activity can be divided into active (talking) and inactive (silence) periods. In a typical full-duplex two-way voice conversation, the duty cycle (talking compared to silence period) of each voice is about 35-40 %. When duty cycle is low, the variable-rate vocoder represents the speech with a lower data rate. The vocoder algorithm uses Coded

Excited Linear Prediction (CELP), and the CDMA specific algorithm is termed QCELP. This added coding efficiency increases CDMA system capacity by a factor related to the ratio of silence to sound intervals in the speech.

In addition to the variable rate encoding process there is also a dynamic threshold adjustment in the vocoder. When the vocoder detects background noise, such as vehicle traffic or wind blowing, the vocoder raises the threshold for noise encoding only active speech and not the background noise.

The vocoder outputs one voice frame every 20 msec. The data in the frame will be either at the full, half, quarter, or eighth rate. The number of bits for each voice frame rate depends on the vocoder used. The vocoders used in CDMA have a maximum voice data rate of 8 kbps and 13 kbps. The vocoder rate of 8 kbps and 13 kbps rates refer to the voice data rate. Each 20 msec voice frame will have overhead bits multiplexed into the frame, which increases the transmit rate. The transmit data rates for the 8 kbps vocoder, with overhead bits, are 9600 bps, 4800 bps, 2400 bps, and 1200 bps. Transmit data rates for the 13 kbps vocoder, with overhead bits, are 14.4 kbps, 7.2 kbps, 3.6 kbps, and 1.8 kbps. Due to the variable rate encoding the average data rate is about 4 kbps and 7 kbps respectfully.

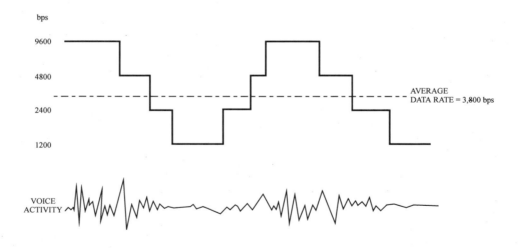

Figure 3.6, Variable Rate Speech Coding

The voice encoding process begins with an analog-to-digital conversion of the user's voice. The analog-to-digital conversion is done at a fixed sample rate of 8,000 samples per second with 8 bits per sample. The resulting data rate is 64 kbps. The vocoder characterizes the speech into parameters that are placed into 20 msec frames.

Figure 3.6 illustrates how the speech coder compression rate varies with speech activity. The speech signal is divided into 20 msec intervals. The bit rate increases and decreases as the speech activity increases and decreases.

At the receiver the frame quality bits (CRC) will be used to aid in determining the frame rate. If the frame quality bits do not check for one data rate, decoding at another data rate will be attempted until the correct data rate is determined. The receiving vocoder then decodes the received speech packet into voice samples.

Radio Channel

A radio channel is typically characterized by path loss (the distance between the MS and cells) and the fading. The physical environment reflects the transmitted signal causing multiple signals of varying strengths. Each signal will travel a different path to the receiver arriving with a time delay when referenced to an earlier arriving signal. At cellular and PCS frequencies these time delays may create deep fades that are 200-300 KHz wide. The fading is a function of the MS position, speed and signal bandwidth. Therefore, a small change in the MS's physical location changes the delay associated with all paths. In the 800 MHz cellular band a MS experiences one fade per second, per mile per hour.

In the 1900 MHz PCS band a MS experiences two fades per second, per mile per hour. Fading is very harmful to the communications channel and requires additional power to overcome.

The CDMA radio channel spreads the signal over a wide 1.23 MHz frequency range, making it less susceptible to radio signal fading that occurs only over a specific narrow frequency range. As a result, radio signal fades affect only a portion of the CDMA signal bandwidth, and most of the information gets through successfully. With only a small portion of information corrupted, digital information transmissions over a CDMA radio channel are relatively robust. [1]

Power Control

Power control is the adjustment of the transmitted power level of the MS or BS transmitter. Power control is very important in CDMA. The system limits both forward link and reverse link traffic channel communication to the minimum transmitter power required for the receiving device to accurately retrieve call data. This reduces the amount of interference that a traffic channel imposes on other channels. Forward link power control is facilitated by the cell site minimizing its transmit power by evaluating signal quality messages from the mobile. Reverse link power control messages are mixed in with the voice data on the traffic channel. They instruct the mobile to either increase or decrease its transmit power.

The power level of the MS is continually adjusted. This allows the MS to overcome the affects of path loss and fading, the power on the FTC, RTC and AC and to provide the correct power level needed to obtain the desired signal quality. The objective of the reverse link power control is to normalize all the signals received at the cell regardless of the MS's position, or propagation loss. Therefore, the signals arrive at the cell with minimum required signal-to-interference ration. Normalizing this helps to maximize the capacity of the CDMA system, in terms of the number of simultaneous users.

When the signal arrives at the cell with too low a received power level the bit error rate will be too high for quality communications. When the signal arrives at the cell with too high a received power level the quality will be good but the interference to other mobiles will be too high. The capacity of the system will be degraded when there is too much interference for other mobiles, sharing the same frequency.

To accommodate this requirement for normalized signal levels CDMA systems precisely control MS power. The power control performs two simultaneous operations: an open loop estimate and a fast closed loop. The open loop estimate is a coarse adjustment and the closed loop is a fine adjustment. The power control system maintains received signals within ± 1 dB (33%) of each other. Demonstrations have also shown that a strong interfering signal reduces the number of users per radio channel in a serving cell site. When interference is too great, mobiles are handed off to another cell.

The open loop and fast closed loop power control is designed for the nominal cases. But there will be occasions for exceptions to the nominal case. For example, a small radius cell need not transmit a high power level as a large radius cell. However, when the MS is a certain distance from a low power cell, it receives a weaker signal than it does from a high power cell. The MS transmits with a higher power than is necessary for the short range. Because of this, each cell transmits a set of parameters designed to make adjustments to the MS's estimate to meet the characteristics of the cell.

Reverse Open Loop Estimate

Path loss is slow in nature and is considered to be the same for both the forward and reverse link. Therefore an open loop estimate is used whenever the MS transmits on the AC or RTC. A mobile's open loop estimate is a coarse adjustment of the RF amplifier and is controlled by feedback from its receiver section. The MS continuously measures the received signal strength from the cell.

The mobile estimates the loss between the MS and cell. Figure 3.7 shows that as the MS moves away from the cell, the received signal strength, at the MS, decreases. When the received signal is strong, the MS reduces its transmit power; conversely, when the mobile's received signal level is weaker, the MS increases its transmit power. The end result is that the signal received at the cell from the mobile remains at about at the same power level regardless of the mobile's distance.

The open loop power control is based primarily on the received power at the MS. This is simple but is not very accurate because of the level of interference at a mobile. To improve this process the open loop power control algorithm was modified to consider the strength of the serving pilot.

Fast Closed Loop

Because the forward and reverse link channels fade differently a fast closed loop power control mechanism is employed to help overcome the fades not apparent to the MS. The cell fine-tunes the mobile's transmit power by sending power control commands to the MS during each 1.25 msec time interval. The command is a single bit sent on a power control sub-channel on the FTC. The location of the power control bit within each 1.25 msec time interval is randomly selected using the PN long code. This command adjusts the transmit power of the MS in 1 dB steps. The adjustment is determined by the received signal strength at the cell. The power control bit communicates the relative change from the previous transmit level, commanding the MS to increase or decrease power from the previous level.

Figure 3.8 illustrates the closed loop power control. As the received signal strength is too high, the power control bit instructs the MS to reduce transmit power. When

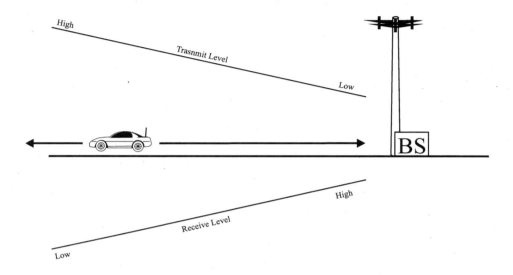

Figure 3.7, CDMA Open Loop RF Power Control

the received signal strength is lower than desired, the power control bit instructs the MS to increase transmit power. The closed loop adjustment range (relative to the open loop) is ± 24 dB minimum. The MS must adjust its output power to within 0.3 dB within 500 msec. The combined open and closed loop adjustments precisely control the received signal strength at the cell. The power control mechanism is analog in nature and has a combined dynamic range of 80 dB [2].

Forward Link Power Control

In conjunction with the reverse link power control there is also a forward link power control. The power for each FTC is dynamically controlled in response to information received from the MS listening to the FTC. In certain locations, the link from the cell to MS may be unusually disadvantaged. This requires the power being transmitted to this MS be increased or it suffers unacceptable signal quality.

When the cell enables FTC power control, the MS reports frame error rate (FER) statistics. When the MS senses an increase in received FER, it reports the increase by sending a message or setting a bit. The message or bit may be sent periodically

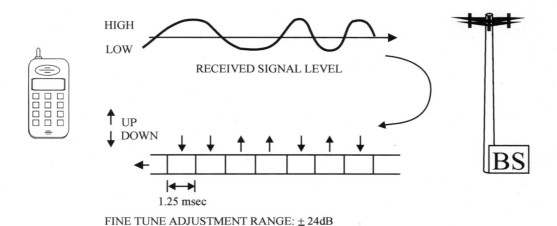

RECEIVED SIGNAL LEVEL

1.25 msec

FINE TUNE ADJUSTMENT RANGE: ± 24dB

Figure 3.8, CDMA Closed Loop RF Power Control

or when the FER reaches a specified threshold. The cell site responds by adjusting its power level that is dedicated to the MS on the FTC. The rate of change in transmit power is slower than used for the MS. The adjustment can be made once every 20 msec.

Average Transmit Power Reduction

CDMA reduces the average power transmitted on FTC by distributing the power allocated to each bit across all the repeated symbols out of the repetition block. Figure 3.9 illustrates how the power required for one symbol of data is now equally distributed to each of the repeated symbols of data.

On the reverse link the MS is able to reduce the average power by employing a burst transmission scheme. The burst transmission reduces the average power over time. To ensure that all mobile stations do not burst at the same moment, each mobile's burst transmission period is randomized.

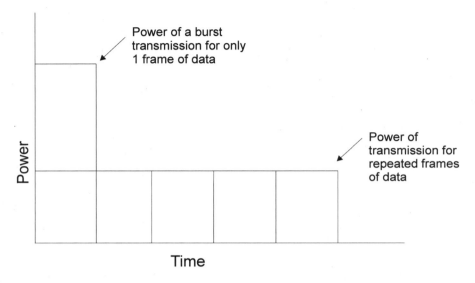

Figure 3.9, CDMA cell Transmit Power

The use of a variable rate vocoder makes this burst transmission possible. The reverse link is the weaker of the two links and is therefore designed to randomly transmit one of the repeated frames of data but at the full power allocated by the power control loops. This transmission burst is 1.25 msec in duration. This period of time corresponds to the frequency of the fast closed loop power control enabling the MS to quickly adjust its transmit power to the match the conditions of the radio channel. Figure 3.10 shows how a 4800 bps channel transmits only 1/2 of the time.

Rake Receiver

In narrowband radio channels, such as those used for analog systems, the existence of multipath signals causes severe fading. With wideband CDMA signals however, the different paths may be discriminated against in the demodulation process. The ability to discriminate greatly reduces the severity of the multipath fading. In a wideband signal, such as with CDMA, that have a 1 MHz PN chip rate, multipath signals greater than one microsecond apart are useful for demodulation. However, multipath signals that are less than one microsecond apart will result in a fading behavior.

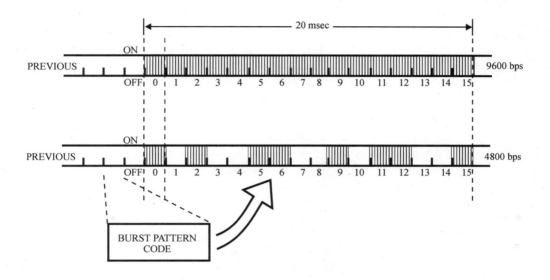

Figure 3.10, CDMA MS Transmit Power Bursts

Both the mobile and cell employ a rake receiver allowing the demodulation of multipath signals. When multipath signals are received at slightly different times a receiver is assigned to the signal. The receiver demodulates the signal and combines it with other weak multipath signals to construct a stronger one. This process is called RAKE reception. The result is better voice quality and fewer dropped calls than would otherwise be possible.

Figure 3.11 illustrates how a multipath signal can be added to the direct signal. The radio channel shows two code sequences. The shaded codes are time delayed because the original signal was reflected and received a few microseconds later. The original signal is decoded by mask #1. The code is shifted in time until it matches the delayed signal. The output of each receiver is coherently combined to produce a better quality signal. Because of CDMA's wideband signal three receivers are used in both the mobile and cells.

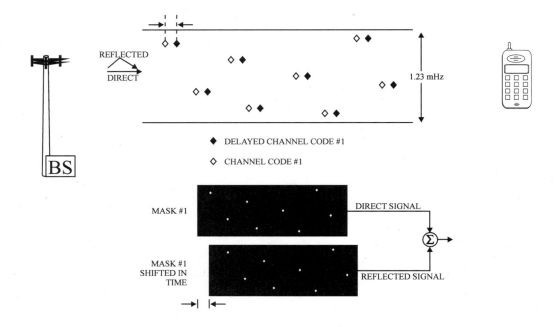

Figure 3.11, CDMA Rake Reception

Soft Handoff

Because each cell uses the same frequency and codes are used to separate users the MS can demodulate two or more FTC simultaneously, this process is referred to as a soft handoff. A soft handoff provides a more reliable and higher quality signal. In AMPS cellular systems a handoff occurs when the cell detects deterioration in signal strength from the MS.

As an AMPS MS approaches a handoff, signal strength may vary abruptly, and the voice is muted for at least 200 milliseconds in order to send control messages, give the MS time to change frequencies, and complete the handoff. This is referred to as a "break-before-make" because the MS must stop transmitting and change to a new frequency. In contrast, CDMA systems use a unique soft handoff while in the Traffic State, in which handoff undetectable.

Soft handoff allows the MS to communicate simultaneously with two or more cells where the best signal quality is selected until the handoff is complete. The standards allow for a soft handoff on up to six FTC at the same time.

The soft handoff is a mobile assisted process. Figure 3.12 illustrates the process of a mobile assisted handoff. The MS measures the PC signal strength from surrounding cells. It then transmits the measurements to the serving cell. The serving cell obtains a WC from the cell #2 and sends it to the MS.

Both the MS and the new cell begin receiving traffic, while the MS continues to receive traffic from the original serving cell. Using multiple cells, with the same frequency, simultaneously during the handoff helps to maintain a much higher average received signal strength at the MS. During soft handoff, each cell sends a vocoded frame of speech to the vocoder where the best frame is selected for decoding. The MS uses its rake receiver to optimally combine the cell signals, even if they do not arrive at the MS synchronized due to varying distances from the two cells.

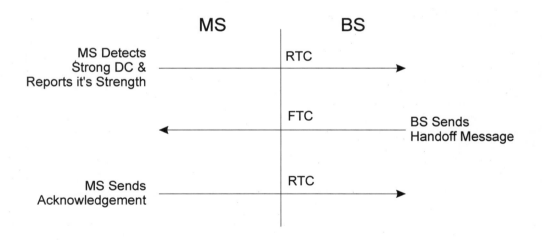

Figure 3.12, MS Assisted Handoff

Sleep Mode

The MS can use a sleep mode when listening to the PgC. The MS goes to sleep (shuts down unnecessary functions) and wakes up on a periodic basis to check messages and for a page. The PgC messages are distributed throughout the PgC and MS directed messages will be inserted into slots where the MS knows to look. Therefore, Discontinuous Reception (DRX) enables MSs, which are not in the Traffic State to power-off non-essential circuitry during periods when pages will not be sent. The sleep cycle may be in multiples of 1.28 sec with the maximum being 163.84 sec. Figure 3.13 illustrates the DRX (sleep mode) process. When the MS registers the Base and MSs select the slot cycle index for the length of the sleep mode.

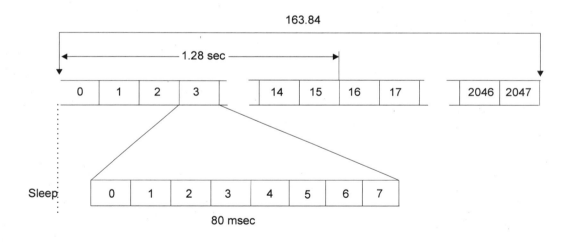

Figure 3.13, Discontinuous Reception (Sleep Mode)

Signaling on the Paging and Access Channels

The CDMA system has been designed with a very flexible signaling and control structure. This allows for additional features and capabilities to be readily added. When a MS is not involved in a call, signaling functions must be provided with the BS. For this purpose, the CDMA system uses the SC, PgC, and AC. Signaling on all channels uses a synchronized bit-oriented protocol. The transmission of the messages is bit synchronized to correspond to the start of the PN short code. The format of the message capsule consists of a message and padding. Padding is used on some channels to make the message fit into a frame.

Framing and Signaling on the Traffic Channel

Signaling is the physical process of transferring control information to and from the MS. Signaling can be either blank-and-burst or dim-and-burst. Blank-and-burst replaces one or more frames of primary traffic data, typically vocoded voice, with signaling data. Dim-and-burst signaling sends control information in the unused portion of a vocoded frame during periods of low speech activity. When a dual mode MS operates on the AMPS channels, signaling is performed using the blank-and-burst method of EIA-553. When operating on the CDMA channel, signaling is sent by blank-and-burst or by dim-and-burst. Variable rate speech coding varies the coding rate allowing both voice and control messages to be sent during each 20 msec frame, permitting both fast or slow dim-and-burst signaling.

Blank & Burst

Blank-and-burst signaling replaces speech data with signal messages. For historical reasons, this is called "in-band signaling." Blank-and-burst message transmissions degrade speech quality because they replace speech frames with signaling information. The quality degradation for only one isolated replacement is almost imperceptible. Figure 3.14 illustrates blank-and-burst signaling.

Dim & Burst

Dim-and-burst inserts control messages when speech activity is low. The QCELP vocoder is designed to produce a lower bit rate when the voice is silent, between syllables and when there is only background noise, for example. As the vocoder changes the bit rate, the lower voice data rate and a portion of the overhead message are multiplexed into a higher rate frame. The 8 kbps vocoder uses only the full rate frame of 9600 bps to multiplex voice and signaling data. The 13 kbps vocoder can use any vocoder frame for voice and signaling.

The number of required frames for sending the message varies according to speech activity, length, and priority of the message. The dim-and-burst messages have no effect on speech quality, but the message requires somewhat more time to be transmitted.

When a vocoded frame has both voice and signaling data a mixed mode bit is set indicating to the receiver there is a mix of voice and signaling data in the frame. Additional flag bits include a Traffic Type and Traffic Mode for the 8 kbps vocoder. For the 13 kbps vocoder there is additional Mixed Mode and Frame Mode Bits. These bits indicate to the receiver whether the message is being sent via blank-and-burst, dim-and-burst, and the proportional mix between voice and signaling data. Figure 3.15 illustrates a dim-and-burst message.

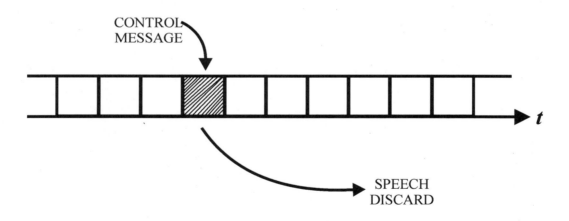

Figure 3.14, CDMA Blank-and-Burst Signaling

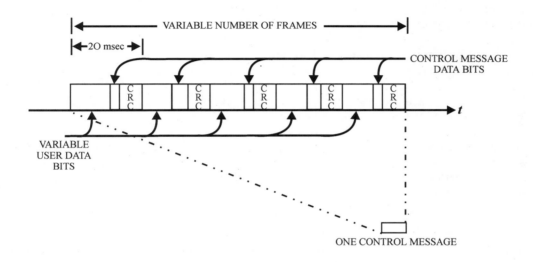

Figure 3.15, CDMA Dim-and-Burst Signaling

Registration

Registration is the process by which a MS notifies the BS of its location. This allows the BS to more efficiently locate the MS for paging. The MS uses the registration to notify the BS of its MINs, PgC slot, Station Class Mark (SCM), and mobile terminated call status.

There is a tradeoff between the rate of registrations and the paging of mobiles. When a MS does not register the BS does not know whether the MS is on, off, or where it is located. Therefore the system must page the entire network placing a heavy load on the PgC. If frequent registrations are used then the system will know where the MS is located with a high degree of accuracy. However, this places a high load on the AC and a moderate load on the PgC, since the system must acknowledge each registration.

The standard defines nine types of registration:

Power-up registration: The MS registers whenever it powers on, when it switches from using an alternate system, switches from using the analog system, or a different frequency band class.

Power-down registration: The MS registers when powering off if it is registered in the current system.

Timer-based registration: The MS registers when a timer expires. The MS timer counts the paging slots to determine time.

Distance-based registration: The MS will register whenever it moves more than a certain distance from where it last registered. The cells send their latitude and longitude, and a distance threshold. The MS determines it has moved a certain distance by computing a distance based on the difference in latitude and longitude between the current cell and the cell where it last registered. If the distance measured exceeds the threshold a registration is sent.

Zone-based registration: The MS registers when it enters a new zone. A MS may be registered in more than one zone. However, the BS controls the maximum number of zones in which a MS may be registered. Parameter-change registration: The MS registers when certain stored parameters change, or when it enters a new system. The parameters that will initiate a registration are a change in MIN, band class, slot cycle, operating mode, or call termination indicators. Ordered registration: The MS registers when instructed to by the BS. Implicit registration: The MS is registered when it makes a call attempt or responds to a message. The BS can infer the MS's location. Traffic Channel registration: When the BS has registration information for a MS that has been assigned to a traffic channel, the BS notifies the MS that it is registered.

The first five methods of registration are called autonomous registration and are initiated in response to an event. The BS can enable or disable any of the various types of registration. It is expected that combinations of registration methods will be the most effective.

Capacity

In the cellular frequency reuse concept, interference is accepted but controlled with the goal of increasing system capacity. CDMA does this effectively because it is inherently a better anti-interference waveform, with its beginnings in military anti-jamming systems. Narrowband modulations are limited in frequency reuse efficiency by the requirement to achieve a C/I ratio of 18 dB. This requires that a frequency used in one cell not be reused in a nearby cell. In CDMA, the wideband channel is reused in every cell.

In CDMA, frequency reuse efficiency is determined by the signal-to-interference ratio that results from all the system users within range. Since the total capacity becomes quite large, the statistics of all the users is more important, than a single user. This means that the net interference to any given signal is the average of all the users' received power times the number of users. As long as the ratio of received signal power to the average noise power density is greater than a threshold value; the channel will provide an acceptable signal quality. An acceptable C/I+C for CDMA in a mobile environment with a 7 dB Eb/No would be "-14 dB".

When deploying a CDMA system where there is existing AMPS requires that consecutive analog channels be cleared the CDMA radio channel. Each CDMA radio channel occupies 1.23 MHz of spectrum (a bandwidth corresponding to about 42 AMPS radio channels). However, one CDMA radio channel typically replaces 2 to 6 AMPS carrier frequencies in a single cell or sector. This is because AMPS radio channels in each cell or sector are placed 7 to 21 channels apart (210 to 630 KHz frequency separation) to allow for n=7, 12, 13, or 21 frequency planning. In an n=7 cell frequency plan with 3 120° sectors, the frequency separation in one sector alone is 21 channels or 630 KHz.

The reverse link capacity of a CDMA system may be ten to twenty times AMPS, depending on the type of deployment. The application and number of sectors deployed in a cell have a large impact in the capacity of an individual cell. The primary factors governing the reverse link capacity include: the processing gain (bandwidth/vocoder rate), voice activity, number of sectors, Eb/No (quality metric), and frequency reuse efficiency.

Sectorization

The capacity of a FDMA or TDMA network is primarily increased by adding cell sites. After the cell sites have been located, the frequency allocation (frequency plan) then determines the maximum number of communication channels for the network. The serving capacity of a CDMA system can be increased by the reduction of interference from other radio carriers. Dividing a CDMA cell into sectors increases the capacity through reduced interference. When adding sectors to a cell a service provider simply adds additional antennas (if necessary) and a radio section for each sector. Current implementations by some infrastructure manufacturers have as many as 24 sectors per cell. The application would determine how many of the sectors would be on a single CDMA radio channel.

Data Transfer

The demand for wireless internet access, e-mail, and fax are growing. Text and multimedia data accounts for 50% of all telephone traffic in the United States. The wireless data market is made of business professionals and field workers. Current technologies have been limited in data capabilities.

TIA/EIA-95 is designed to satisfy the growing data requirements with an optional Medium Data Rate (MDR). The MDR allows up to eight FTC to be used together. There is a fundamental channel and supplemental channels defined. The fundamental channel will be used to carry primary (voice), secondary, and or signaling traffic. Signaling will only occur on the fundamental channel. Supplemental channels may carry primary (voice) or secondary data, but not both. The supplemental channel must be the same rate as the fundamental channel and will only operate at the full data rate.

Enhanced 911

In 1968 the emergency "911" started in America. Since then it has spread across the nation and over 100 million calls are made each year. These calls are answered at a central facility where police, fire, and medical professionals are dispatched. When a call is placed to a facility capable of Enhanced 911 (E911) the local phone company transmits the callers telephone number and telephone location to operator at the facility.

Current wireless carriers provide basic 911 service and not E911. Because of the mobility of the caller the position is not always possible to know. The FCC has mandated that carriers provide the capability to identify the location of a MS making a 911 call, within a radius of no more than 125 meters in 67% of all cases.

To facilitate this requirement a Power Up Function (PUF) has been developed. The PUF defines the technique used to increase the MS transmit power in a very controlled manner in order to aid MS visibility. The BS controls the PUF process. PUF is a triangulation technique used to locate the MS when several cell sites are able to receive the MS signal. The MS briefly increases its power until multiple cell sites are able to receive the signal and provide to the E911 operator the MS location.

Handoffs

The current CDMA standard defines three MS states when a handoff may occur. These states are the Idle State, Access State, and Traffic State.

Idle Handoff

The idle handoff will occur in the Idle State and will take place when a MS has moved from the coverage area of one cell into the coverage area of another cell. When the MS detects that a PC from a new cell is twice as strong (3 dB) than the current PC; the MS starts listening to the PgC of the new cell.

Access Handoff

Access Handoffs are similar to Idle Handoffs. All are hard handoffs from one PgC to another. The Access Handoffs may occur before the MS begins sending access probes, during access probes, and after receiving an acknowledgment to a probe.

The Access Entry Handoff allows MS to perform an Idle Handoff to the best cell. This handoff would occur just after entering the Access State but before updating the information from the PgC.

After the MS begins making access probes a new and stronger pilot may provide a better chance of service. The MS can then perform a Access Probe Handoff to prevent the access attempt from failing.

After the MS has received an acknowledgment to a probe the access attempt is complete. A handoff to a new and stronger pilot may be possible and necessary to prevent an access failure.

Soft Handoff

A soft handoff is when new cells or a sector of the existing cell communicates on a FTC with the MS without interrupting the FTC of the old cell or sector. The BS can direct the MS to perform a soft handoff only when the FTC of the new cell or sector has the same frequency, band classes, and frame offset.

During the soft handoff while in the Traffic State, the MS receives a list of neighboring cells from the PgC. This list of neighboring sites is a list of PN offsets for neighboring cells' pilot channels. These PN offsets are the most likely candidates for handoff. When the PC of new cell has sufficient energy at the MS for handoff, the MS requests simultaneous transmission from that channel. The system then assigns the new channel to transmit simultaneously from both the current cell and the new cell. The MS decodes both channels (different codes on the same frequency) and optimally combining the received information. When the signal strength of the original channel falls below a threshold, the MS requests release of the original channel ending the communication with the cell. The MS is capable of simultaneously communicating with two or more cells at the same time.

A soft handoff is initiated when a pilot strength goes above the threshold for adding a pilot (the threshold is called T_ADD). The pilot will be considered usable until it drops below and stays below a defined drop threshold (the threshold is called T_DROP) for more than a defined amount of time (TT_DROP). The soft handoff is based on the threshold level of the received pilot. Although this is a simple method for determining when to perform a handoff to a new cell it is possible for a MS to obtain more FTC than would be necessary, thus adding to the level of interference. Therefore dynamic thresholds parameters have been added to the soft handoff process. These dynamic thresholds are used to trigger handoff based on a comparison of pilots to the combined pilot energy. To aid the MS, in the search, parameters and process for sending messages, there are four pilot sets defined.

Active Set: All pilots associated with assigned FTC.
Candidate Set: Pilots with sufficient signal strength to indicate that a FTC could be demodulated but have not been assigned to a FTC.
Neighbor Set: Pilots that are likely candidates for handoff.
Remain Set: The set of all possible pilots in the current system, excluding the pilots in the Neighbor Set, Candidate Set, and Active Set.

Hard Handoff

A CDMA-to-CDMA hard handoff occurs when the BS directs the MS to transition between disjointed sets of networks, different frequency assignments, different frame offsets, or different band classes.

A CDMA-to-Analog hard handoff occurs when the BS commands a MS to change from a FTC to an 800 MHz analog voice channel. The hard handoff can occur on a CDMA system operating in the cellular or PCS bands. A hard handoff from PCS band to cellular band analog mode requires the mobile to support the analog-operating mode. Since CDMA is a migration from AMPS the need for AMPS to CDMA hard handoff is not necessary. Because there are no AMPS to CDMA handoffs, the signaling format of the AMPS voice channel is unchanged.

Call Processing

During normal operation, a CDMA network sends, receives and processes messages that operate the mobile phone and setup equipment in the network to allow communication. The sequence of events that are required to set up and process voice calls is referred to as call processing.

Many of the call processing steps are repetitive. When grouped together, call processing sequences are called tasks. Key call processing tasks include initialization, idle, access, traffic channel communication, power control, handoffs, and call termination. All of these individual steps are necessary for a mobile to operate on the network. During the processing of these tasks, the mobile radio is in a state of operation. Figure 3.16 illustrates the basic flow from each state in the MS call processing.

Initialization State

Initialization is the process of obtaining system information (parameters) that instruct the mobile radio on how to communicate with the system. The initialization process is triggered when the mobile radio is first turned on. One of the first steps in the initialization mode is for the mobile radio to selects a system to use (CDMA or analog). If the selected system is AMPS then the MS follows the analog acquisition procedures.

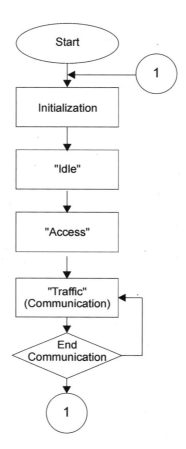

Figure 3.16, MS States

If the selected system is a CDMA system, the MS proceeds to acquire and synchro-nize with the system. The MS searches for a pilot channel by aligning its PN short code with a received PN short code. Once a valid pilot channel is found, the mobile synchronizes with it. The MS has fifteen seconds to find and acquire a PC. If the MS does not find a CDMA signal it may decide to search for an AMPS (analog system) control channel. When it finds the PC, the MS switches to walsh code (WC) one and looks for the start of the SC message. The SC contains information about system time and the PN codes needed to synchronize its PN codes. After decoding the SC it aligns its timing to that of the BS. Figure 3.17 illustrates the flow in the Initialization State.

Idle State

After initialization, the MS enters the Idle State. During the idle state, the MS is waiting to receive or originate calls and monitoring the channel quality of active radio channels. During the idle state, the mobile station regularly decodes the PgC to obtain system configuration parameters, access parameters, and list of neighboring sites.

After the MS has acquired key information about the system, it may be allowed to enter into a sleep mode to conserve battery life. When the MS goes into a sleep

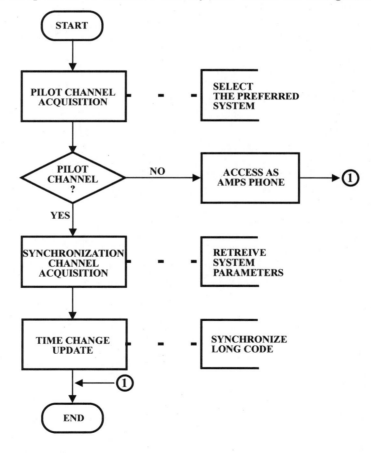

Figure 3.17, CDMA Initialization State

mode, non-essential operations are turned off for fixed periods of time. Because both the BS and MS know the length of the sleep time, the MS knows the periods when messages will be sent for it on the PgC.

To ensure the mobile radio will communicate with the best possible radio channel, the MS will regularly monitor the radio channel quality of its serving and other active radio channels. Neighboring cell use different PN short codes to identify their signals, so the MS will simultaneously monitor the PN offset of neighboring cell. If

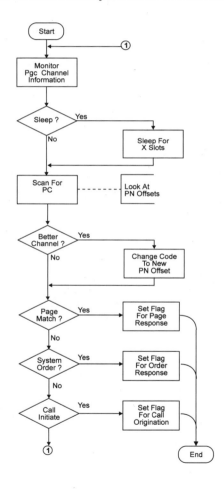

Figure 3.18, CDMA Idle State

the MS detects a PC with a higher power the MS begins monitoring the PgC associated with the stronger PC. The MS monitors the PgC for messages and updates. The MS will enter the Access Sate when it receives a mobile directed message.

Although the MS is not directly communicating when it is operating in the idle mode, the mobile radio is essentially locked to a specific base station. This allows messages only to be sent to the active base station where the mobile is operating. When a mobile radio determines that another base station is a better candidate for communicating, the MS will attempt to signal to the new base station indicating that it will be operating in its coverage area. If the BS has the resources available (all the channels may be busy), it may accept the mobile radio's request to be transferred to the new base station. This process is called Idle handoff. The idle handoff is a mobile's way of updating the network with its current position, and updating itself with the most current configuration information. When an idle mobile moves from one sector's coverage area into another sector's coverage area, it performs an idle handoff. The mobile monitors the new sector's paging channel and stores new network configuration information. Figure 3.18 illustrates the Idle State processing.

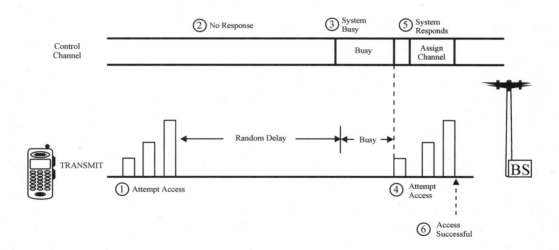

Figure 3.19, CDMA Access Probes

Access State

In the Access State the MS sends messages in response to mobile directed messages, registration, or call origination. The MS randomly attempts to access the system.

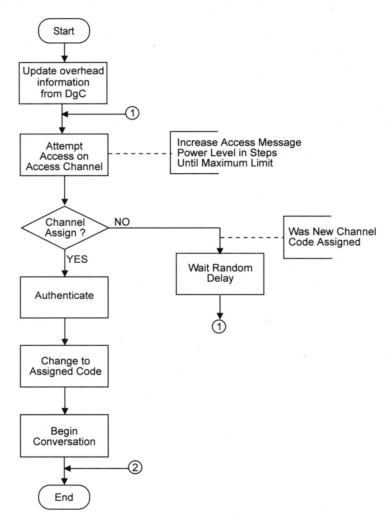

Figure 3.20, CDMA Access State

Access to the system is gained when a response is received on the PgC. Multiple MSs, associated with a particular PgC, may simultaneously try to use the same AC causing a collision at the cell. To mitigate this problem a random-access protocol is developed to avoid collisions. The collision avoidance protocol consists of access class groups; gradual increase in access request power levels, random time delays, and maximum number of automatic access attempts.

Figure 3.19 illustrates AC probing. The MS sends access probes of increasing power level. The probes continue until the maximum number have been exceeded, or the MSs power limit has been reached, or it receives an acknowledgment. When a probe is unsuccessful the MS waits for a random period before making another attempt.

Figure 3.20 illustrates the Access State processing. Before an access attempt the MS randomly selects both an AC and a time offset of the long PN code used for spreading. A MS will transmit successive probes until a response is received or it exceeds the allowed access attempts. Between each set of probes and each attempt the MS will wait a random period to time. The time between probes and attempts can be varied for different types of transmissions and access classes of with priorities for emergencies and maintenance. MSs responding to a message on the PgC do not have to wait to begin their access attempts.

When a user places a call, resources may not be available due to all channel elements being in service or no available transmitter power. Priority Access Channel Assignment (PACA) is a service where priority is given to MS originations. When required FTC resources are not available the call is placed in a queue. Originations in a queue are given priority over newly arriving originations. When resources become available the next call in the queue is set-up with the MS.

Traffic State

When a mobile radio is transferring user information between mobiles and cell sites, this is called the traffic state. This information could be a voice conversation or other transmitted data, originating from another mobile in the network, another mobile in another network, or from a land line. In Traffic State the MS sends voice and control data on the RTC while receiving voice and control data on the FTC.

The MS continually receives data on the FTC. When the MS is in soft handoff it receives the same data on different FTC from multiple cell or sectors. The received FTCs are demodulated and coherently combined. Any control message is extracted from the data and processed. Several frames will be needed to receive the message.

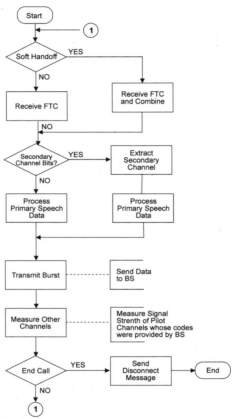

Figure 3.21, CDMA Traffic State

The voice data is extracted and sent to the vocoder for processing. The MS sends voice and signaling data in a 1.25 msec bursts on its RTC. The MS will continually search for other strong PC that may be useful for traffic. After the Traffic State the MS enters the Initialization State. Figure 3.21 illustrates the Traffic State processing.

Authentication

The authentication process verifies the identity of a mobile, to the network. The mobile does this by transmitting identifying information to a cell site. The cell site compares the received information with the information stored in the network records for the particular mobile. If they match, the mobile is allowed to access the network's services. If they do not match, network access is denied. This process effectively protects the network from fraud.

Call Termination

Call termination occurs at the end of a voice conversation, and is initiated by either the mobile or the network. When the mobile radio terminates a call (usually by pressing the end button), it sends a call termination message to the cell site, stops transmitting on the traffic channel, and starts the initialization process. If the network terminates a call, when the other party hangs up the phone, the cell site sends a call termination message, the mobile stops transmitting on the traffic channel and starts the initialization process.

MS Originated Call

When a mobile originates a call, it sends a system access message on the access channel while monitoring the paging channel for responses. Once system access is complete, the mobile enters the authentication process. This prevents any unauthorized mobiles from accessing the network services. If the authentication is successful, a traffic channel is assigned, and voice communication can occur between the mobile and the called party. Many transmit power changes and handoffs can occur while on a traffic channel. When the voice conversation is over, the call is terminated and the mobile starts the initialization process.

Figure 3.22 illustrates the message flow, code channel usage and timing for a MS originated call. This diagram shows that after the MS detects that a user has initiated a call (by dialing and pressing SEND), the MS a message on the AC requesting service (step 1). The cell assigns the WC to the transmitter section for the FTC and then receivers to the RTC PN long code. It then begins sending null traffic on the FTC and sends a PgC message containing the WC assigning the MS to the FTC

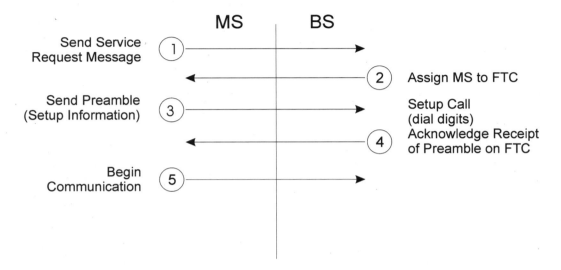

Figure 3.22, MS Originated Call

(step 2). The MS configures its transmitter section for the RTC and its receivers for the FTC, begins decoding the null traffic on the FTC and starts sending a preamble on the RTC (step 4). The cell uses the FTC to acknowledge the preamble (step 5). The MS receives the acknowledgment and starts transmitting traffic (step 6).

BS Originated Call

When a call has been placed to a MS, it receives a message on the paging channel, informing it of an incoming call (paging message). The mobile sends a response on the access channel and a traffic channel is then assigned. Many transmit power changes and handoffs can occur while on a traffic channel. When the voice conversation is over, the call is terminated and the mobile starts the initialization process.

Figure 3.23 illustrates the paging process. The MS listens for pages on the PgC. Initially (step 1), the MS receives a page on the PgC. The MS sends an acknowledgment on the AC (step 2). The cell receives the acknowledgment, configures the transmitter to the assigned WC for the FTC and the receiver to the RTC PN long code. It then begins sending null traffic on the FTC and sends a PgC message containing the WC assigned to the FTC (step 3). The MS configures its transmitter section to the RTC PN long code and its receivers to the WC assigned to the FTC. It

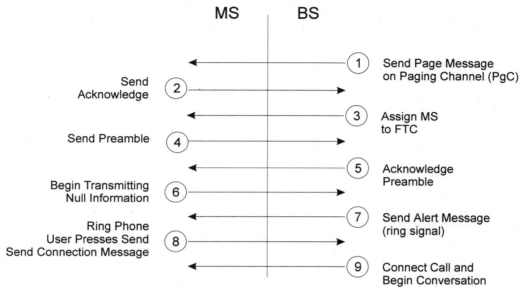

Figure 3.23, BS Originated Call

then begins decoding the null traffic on the FTC and starts sending a preamble on the RTC (step 4). The cell uses the FTC to acknowledge the preamble sent on the RTC (step 5). The MS receives the acknowledgement and begins sending null traffic on the RTC (step 6). The cell then sends an alert message on the FTC for a ring tone and calling number indicator (step 7). The mobile acknowledges the message by ringing the handset and displaying the calling number information. When the user answers a connection message is sent on the RTC (step 8). The BS acknowledges the connection message and sends voice data (step 9).

Handoff

A handoff will occur as the MS moves from one location to another. As it moves away from the cell currently handling its call the PC will grow weaker. At the same time the PC from an adjacent cell will grow stronger.

Figure 3.24 illustrates the CDMA soft handoff processing. When the PC from cell 2 exceeds the T_ADD threshold, the MS sends a message on the RTC to cell 1 reporting the strength of the PC (step 1). Cell 2 begins sending traffic on the FTC and acquires the RTC (step 2). Cells 1 and 2 send a message instructing the MS to use

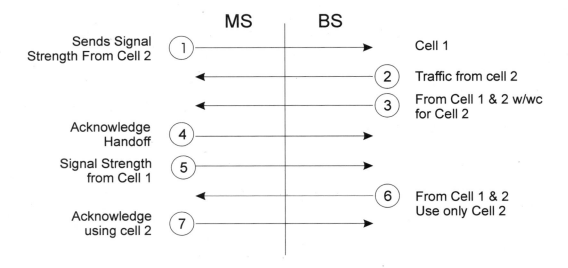

Figure 3.24, CDMA Soft Handoff Processing

cell 1 and 2 (step 3). The message sent to the MS contains the WC for the FTC from cell 2. The MS acquires cell 2 and sends on the RTC acknowledging the message and it has completed the handoff (step 4). When the PC from cell 1 grows weak and stays below the T_DROP threshold, the MS sends a message on the RTC to the cell reporting the strength of the PC (step 5).

Both cells 1 and 2 send a message to the MS on the FTC instructing the MS to use only cell 2 (step 6). The MS stops combining the FTC and only uses cell 2. It then sends on the RTC an acknowledging message (step 7). The MS may use as many as six FTC simultaneously.

References:

1. Industry expert, personal interview, 3 April 1999.
2. ANSI J-STD-008; par. 2.1.2.3.1

Chapter 4

CDMA Mobile Radios

Mobile stations (MS) are the link between the customer and the wireless network. Mobile stations must provide a method for the customer to control the phone and display the operating status. Phones must also sample and process audio signals, transmit and receive radio signals via an RF section and obtain energy to operate the complex electronics, typically from a battery. The basic parts of a mobile phone include:

- Man-machine interface (display and keypad)
- Radio frequency section
- Signal processing (audio and logic)
- Power supply/battery

Mobile stations may be mobile radios mounted in motor vehicles, transportable radios (mobile radios configured with batteries for out-of-the-car use), or self-contained portable units. The official name of a mobile telephone from the CDMA industry standard specification is "Mobile Station".

In addition to the key assemblies contained in a mobile phone, accessories must be available that include battery chargers, hands free assemblies and data adapters. These accessories must work together with the mobile phone as a system. For example, when a portable mobile phone is connected to a hands free accessory, the mobile phone must sense that the accessory is connected, disable its microphone and speaker, and route the hands free accessory microphone and speaker to the signal processing section.

CDMA systems basically use a single type of digital radio channel to provide many types of radio services. Optionally, CDMA systems may also use analog AMPS radio channels if digital radio channels are not available.

There are several different frequencies used by CDMA phones. Some CDMA phones have the capability to operate on dual frequencies (800 MHz and 1900 MHz). Figure 4.1 illustrates the functional sections of a single frequency band CDMA mobile phone.

Man-Machine Interface (MMI)

Customers control and receive status information from their mobile phone via the man-machine interface (MMI). This interface consists of audio input (microphone) and output (speaker), display device, keypad and an accessory connector to allow

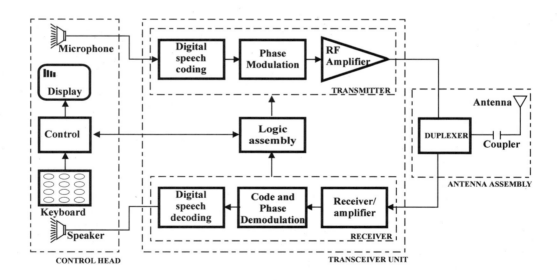

Figure 4.1, CDMA Mobile Telephone Functional Block Diagram

optical devices to be connected to the mobile phone. The mobile phone's software coordinates all these MMI assemblies.

Audio Interface

The audio interface assembly consists of a speaker and microphone that allow customers to talk and listen on their mobile phone. While the audio assemblies are located in a handset, they can be temporarily disabled and replaced by a hands free accessory.

The microphone in small portable telephones is very sensitive. It allows the mobile phone to detect normal conversation even when the microphone is not placed directly in front of the speaker's mouth. This is especially important for the micro-portable phones that have a length smaller that the distance between a customer's ear and mouth.

Display

Mobile phones typically have a display assembly that allows the customer to see dialed digits, status information, stored information, messages and call status information such as radio signal strength. For CDMA mobile phones, dialed digits are displayed and can be changed before the call is initiated. This is known as pre-origination dialing.

Status indicators provide the customer with key information about their phones' operation. These status indicators typically include a "Ready" indication, RF signal strength, battery level, call indication and others. Some of these indicators are icons or text messages. Because mobile phones must find available service prior to requested service, the display may indicate "Wait" while the phone is searching for an available system. To allow the customer to know if they are in an area that has good radio signal strength that is sufficient to initiate or maintain a call, a RF signal level indicator is typically provided. For portable phones, a battery level indicator may be provided indicating the remaining capacity of the battery that may be available to initiate or finish a call.

Most mobile phones have the capability to store and manipulate small amounts of information in an electronic phonebook. In addition to storing phone numbers, some models allow the storage of a nametag along with the number. Because many mobile phones can only display 8 to 12 characters across, nametags are typically limited to only a few letters.

The CDMA system is capable of sending text messages to mobile phones. Mobile phones have several creative approaches to displaying these alphanumeric messages. Some phones show messages in 'pages' one screen at a time. Other phones use a technique known as marqueeing in which a message is scrolled across the screen. This allows the mobile phone that can display only a few characters per line, to display lengthy messages to the customer.

Keypad

The keypad allows the customer to dial phone numbers, answer incoming calls, enter name tags into the phones memory or, in some cases, use the phone as a remote control device via the cellular system. While a keypad is typically used in mobile phones, the keypad may sometimes be replaced by an automatic dialer (auto-security) or by a voice recognition unit.

The layout and design of keypads vary from manufacturer to manufacturer. A typical keypad will contain keys for the numbers 0 to 9, the * and # keys, volume keys and a few keys to control the user functions. In addition, keypads will also contain a SEND and END button that starts and ends calls. Special function keys may also be included for speed dialing and menu features. Customers can typically access special phone options (such as the type of ringing sound or volume) via the keypad.

Accessory Interface

There are several accessories for mobile phones that can typically be attached via an accessory connector (plug). The accessory connector typically provides control lines (for dialing and display information), audio lines (input and output), antenna connection, and power lines (input and output from the battery). Accessory connections are typically proprietary and an industry standard accessory interface connection has not been developed for mobile telephones. Each manufacturer, and often each model, will have a unique accessory interface. The types of accessories vary from passive devices like external antennas to active devices such as a hands-free speaker cradle, computer controlled devices and various power supply options. Figure 4.2 shows a typical accessory connection.

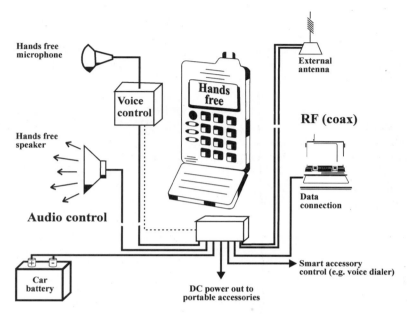

Figure 4.2, Typical Accessory Connection

Radio Frequency Section

The mobile phone's radio frequency (RF) section consists of a transmitter, receiver, and antenna assemblies. The transmitter converts low level audio signals to modulated shifts in the RF carrier frequency. The receiver amplifies and demodulates low level RF signals into their original audio form. The antenna section converts RF energy to and from electromagnetic signals.

Transmitter

The transmitter section contains a modulator, a frequency synthesizer, and a RF amplifier. The modulator converts audio signals to low-level radio frequency modulated radio signals on the assigned channel. A frequency synthesizer creates the specific RF frequency the cellular phone will use to transmit the RF signal. The RF amplifier boosts the signal to a level necessary to be received by the Base Station.

The transmitter is capable of adjusting its transmitted power up and down in 1 dB steps dependent on its received signal strength (course adjustment) and commands it receives from its serving base station (fine adjustment). This allows the mobile phone to only transmit the necessary power level to be received at the serving base station and this reduces interference to nearby base stations that may be operating on the same frequency.

The CDMA digital radio channel phase modulation that requires the use of linear (precision) amplifiers. The frequency accuracy for CDMA mobile phone transmitters requires the use of more precise frequency control than is used for analog phones. To maintain accurate frequency control, the frequency synthesizer (signal generators) is locked to the incoming radio signal of the base station. If the mobile phone has the capability for dual bands of frequencies, additional filters must be used for the transmitter section to allow for both 800 MHz and 1900 MHz channels.

Receiver

The mobile phone's receiver section contains a bandpass filter, low level RF amplifier, RF mixer, and a demodulator. RF signals, from the antenna, are first filtered to eliminate radio signals that are not in the cellular band (such as television signals). The remaining signals are sent to a low-level RF amplifier and are routed to a RF mixer assembly. The mixer converts the frequency to a lower frequency that is either directly provided to a demodulator, or sent to another mixer to convert the frequency even lower (dual stage converter). The demodulator converts the proportional frequency or phase changes into low level analog or digital signals. If the mobile phone has the capability for dual bands of frequencies, additional filters and a high frequency mixer must be used to allow for both 800 MHz and 1900 MHz channels.

Antenna

An antenna section converts electrical energy into electromagnetic energy. Although antennas are passive devices (no signal amplification possible), they can provide a signal gain by focusing the transmission in a desired direction.

The antenna may be an integral part of the mobile phone (in a handheld portable phone) or may be externally mounted (on the top of a car). Antennas can have a gain where energy is focused into a beamwidth area. This focused energy gives the ability to communicate over greater distances, but as the angle of the antenna changes, the direction of the beam also changes, reducing performance. For example, car-

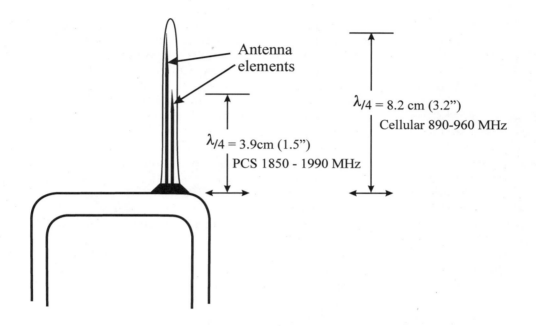

Figure 4.3, Dual Band Antenna

mounted antennas that have been tilted to match the style lines of the automobile often result in extremely poor performance.

If a mobile phone is connected to an external antenna, the RF cabling that connects the transmitter to the antenna adds losses, which reduce the performance of the antenna assembly. This loss ranges from approximately .01 to .1 dB per foot of cable, depending on the type of cable and the RF frequency.

Typical antennas are a quarter wavelength long. If the mobile phone has the capability for dual bands of frequencies, the antenna must be designed to operate at both 800 MHz and 1900 MHz. Figure 4.3 shows an example of a dual band antenna.

Signal Processing

Signal processing is the manipulation of electrical signals from one form to another. Signal processing can be analog signal processing (e.g. high frequency filtering) or digital signal processing (e.g. data compression). A majority of the signal processing for CDMA phones is digital. To process the digital signals, these mobile phones typically use high speed digital signal processors (DSPs) or application specific integrated circuits (ASICs). CDMA phones require over 60 million instructions per second (MIPS), compared to the less than 1 MIPS system processing required by analog cellular phones.

The power consumption of high speed digital signal processors (DSPs) are roughly proportional to the operating voltage and processing speed of the DSP. The first DSPs consumed well over 20 to 30 mWatts per MIP, which was a challenge for extended battery life in portable mobile phones. The newest commercial DSPs technologies consume less than 5 mWatts per million instructions per second (MIP).

Speech Coding

Speech coding is the process of analyzing and compressing a digitized audio signal, transmitting that compressed digital signal to another point, and decoding the compressed signal to recreate the original (or approximate of the original) signal. A significant portion of the digital signal processing that is used in CDMA phones involves digital speech compression and expansion that is called "speech coding."

The CDMA speech coding process involves the conversion of analog (audio) signals into digital signals (analog to digital conversion). The digitized voice signal is then further processed by a speech-coding program to create a characteristic representation of the original voice signal (key parameters). After error protection is added to the compressed digital signal, this information is sent via the radio channel to be transmitted. When it is received, it is recreated to the original analog signal by decoding the information using a speech-decoding program. The CDMA system can use various types of speech coding to maintain system backward compatibility and at the same time, offering enhanced voice quality to customers.

Channel Coding

The process of adding error protection, detection bits and multiplexing control signals with the transmitted information is called channel coding. Error protection and detection bits (they may be the same bits) are used to detect and correct errors that occur on the radio channel during transmission. The output of the speech coder is encoded with additional error protection and detection bits, according to the channel coding rules for its particular specification. This extra information allows the receiver to determine if distortion from the radio transmission has caused errors in the received signal. Control signals such as power control, timing advances, and frequency handoff must also be merged into the digital information to be transmitted. The control information may have a more reliable type of error protection and detection process that is different than the speech data. This is because control messages are more important to the operation of the mobile phone than voice signals. The tradeoff for added error protection and detection bits is the reduced amount of data that is available for voice signals or control messages. The ability to detect and correct errors is a big advantage of digital coding formats over analog formats but it does come at the cost of the additional data required.

Audio Processing

In addition to the digital signal processing for speech coding and channel coding, a digital mobile phone can do other audio processing to enhance its overall quality. Audio processing may include detecting speech among background noise, noise cancellation or echo cancellation.

Echo is a particular problem for audio signals in digital systems. Echo can be introduced by the delay involved in the speech compression algorithm or through normal speakerphone operation. The echo signals can be removed by sampling the audio signal in brief time periods and looking for previous audio signal patterns. If the echo canceller finds a matched signal, it is subtracted, thus removing the echo. While this sounds simple, there may be several sources and levels of echo and they may change over the duration of the call that complicates this process.

Logic Control Section

The logic control section usually contains a simple microprocessor or microprocessor section stored in a portion of an ASIC. The logic section coordinates the overall operation of the transmitter and receiver sections by extracting, processing and inserting control messages. The logic control section operates from a program that is stored in the mobile phone's memory.

There are various types of memory storage that are used in CDMA mobile phones. Part of the memory holds the operating software for the logic control section. Some phones use flash (erasable) memory to allow the upgrading of this operating software to allow software correction or the addition of new features. Typical amounts of memory for CDMA phones are 1 megabyte or more. While the operating software is typically loaded into the phone at the factory, some mobile phones have the capability to have their memory updated in the field. This is discussed later in the software download section.

Read only memory (ROM) is used to hold information that should not be changed in the phone such as the startup processing procedures. Random access memory (RAM) is used to hold the temporary information (such as channel number and system identifier). Flash memory is typically used to hold the operating program and user information (such as names and stored phone numbers).

Subscriber Identity

Each mobile phone must contain identification information to provide a unique identification to wireless systems. CDMA mobile phones have several unique codes.

The most basic form of identification is stored in the number assignment module (NAM). The NAM contains information specific to a subscriber, such as its telephone number, unique serial number and home system identification information. The information contained in its NAM is used to identify the phone to a cell site and mobile switching center (MSC).

The unique identification number for a CDMA mobile telephone customer includes the mobile identification number (MIN), International Mobile Station Identity (IMSI) and the electronic serial number (ESN). The MIN is the mobile telephones' phone number when operating in the analog mode. When operating in the digital mode, the IMSI is the phone number of the mobile radio. The IMSI consists of up to 15 numerical characters (0-9). The first three digits of the IMSI are the mobile country code (MCC), and the remaining digits are the national mobile station identity (NMSI). The NMSI consists of the mobile network code (MNC) and the mobile station identification number (MSIN).

The ESN is a non-changeable number that is stored in the telephone at the time of manufacturing. There should be only one unique ESN for each mobile phone that is manufactured. The ESN is composed of 32 bits. The eight most-significant bits (bit 31 through bit 24) represent the manufacturer's (MFR) Code. Bits 23 through 18 are reserved (initially all zero). Bits 17 through 0 are uniquely assigned by each manufacturer. When a manufacturer has assigned all the possible combinations of their assigned serial numbers, the manufacturer may submit a request to the FCC for the allocation of the next sequential binary number within bits 23 through 18.

Power Supply

There are a variety of power supply options that include batteries, converters and chargers. There are typically several sizes (capacities) of batteries for each model of mobile phone made by a manufacturer. The capacity of the battery varies dependent on its size and battery technology. Most mobile phone models also have available a "Battery Eliminator" that is used to directly connect a mobile phone to a cigarette lighter in the car. When connected to the cigarette lighter, battery eliminators can also to charge the battery of the phone.

There are several types of batteries used in mobile phones: Alkaline, Nickel Cadmium (NiCd), Nickel Metal Hydride (NiMH) and Lithium Ion (Li-Ion). A new type of battery technology, Zinc Air, is being tested that has increased energy storage capacity.

With the introduction of portable cellular phones in the mid 1980s, battery technology became one of the key technologies for users of cellular phones. In the mid 1990s, over 80% of all cellular phones sold were portable or transportable models rather than fixed installation car phones. Battery technology is a key factor in determining portable phones' size, talk time and standby time.

Batteries are categorized as primary or secondary. Primary batteries must be disposed of once they have been discharged while secondary batteries can be discharged and recharged for several cycles. Primary cells (disposable batteries) which include Carbon, Alkaline, and Lithium have a limited use in cellular mobile phones.

Batteries

Alkaline batteries are disposable batteries that are rarely used for mobile phones. This is because although alkaline batteries are readily available, they must be replaced after several hours of use and are more expensive than the cost of a rechargeable a battery. However, disposable batteries have the advantages of a very long shelf life and no need for a charging system. Alkaline batteries are generally not well suited for use in cellular phones because the high current demands of the phones in transmit mode limits the useful life of the battery.

NiCd batteries are rechargeable batteries that are constructed of two metal plates made of nickel and cadmium placed in a chemical solution. A NiCd cell can typically be cycled (charged and discharged) 500 to 1000 times and is capable of providing high power (current) demands required by the radio transmitter sections of portable mobile phones. While NiCd cells are available in many standard cell sizes such as AAA and AA, the battery packs used in cellular phones are typically uniquely designed for particular models of mobile phones. Some NiCd batteries can develop a memory of their charging and discharging cycles and their useful life can be considerably shortened if they are not correctly discharged. This is known as the "memory effect," where the battery remembers a certain charge level and won't provide

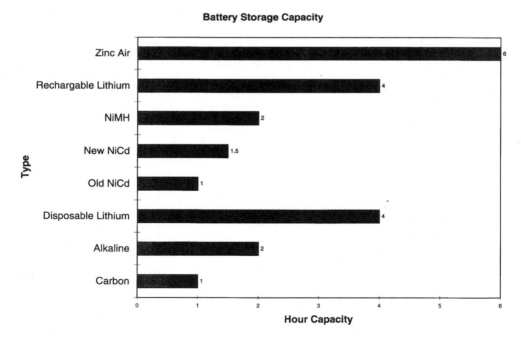

Figure 4.4, Battery Storage Capacity

more energy even if completely charged. Newer NiCd batteries use new designs that reduce the "memory effect."

NiMH batteries are rechargeable batteries that use a hydrogen adsorbing metal electrode instead of the cadmium plate. NiMH batteries can provide up to 30% more capacity than a similarly sized NiCd battery. However, for the same energy and weight performance, NiMH batteries cost about twice as much as NiCd batteries.

Li-Ion batteries are either disposable or rechargeable batteries. Li-Ion is the newest technology that is being used in mobile phones. They provide increased capacity versus weight and size. A typical Li-Ion cell provides 3.6 volts versus 1.2 volts for NiCd and NiMH cells. This means that only one-third the number of cells is needed to provide the same voltage. Figure 4.4 shows the relative capacity of different battery types.

Battery Chargers

There are two types of battery chargers: trickle and rapid charge. A trickle charger will slowly charge up a battery by only allowing a small amount of current to be sent to the mobile phone. The battery charger may also be used to keep a charged battery at full capacity if the mobile phone is regularly connected to an external power source (such as a car's cigarette lighter socket). Rapid chargers allow a large amount of current to be sent to the battery to fully charge it as soon as possible. The limitation on the rate of charging is often the amount of heat generated. That is, the larger the amount of current sent to the battery the larger the amount of heat.

The charging process is controlled by either the phone itself or by circuits in the charging device. For some batteries, rapid charging reduces the amount of charge and discharge cycles. A charger will charge for a period of time, until a voltage transient occurs (called a knee voltage) and checks for temperature of the battery. The full charge is indicated by a couple of different conditions. Either the temperature of the battery has reached a level where the charging must be turned off (the voltage level will reach its peak value for that battery type) or the voltage level will stop increasing. Most chargers will then enter a trickle charge mode to keep the battery fully charged. Some chargers for NiCd batteries discharge the battery before charging to reduce the memory effect. This is called battery reconditioning.

Standby and Talk Time

CDMA technology provides unique tools for decreasing the phone power consumption and increasing the standby time of batteries. The operation of the digital control channel is such that phones do not have to receive the control channel information continuously like in an analog system. Since the phone and system are synchronized together, the phone can power off some of its circuitry while waiting between messages. This means that typically, only a few messages of information has to be read on the control channel each second - drastically increasing the battery life.

Accessories

Accessories are optional devices that may be connected to a mobile phone to increase their functionality. Accessory devices include hands free speakerphone, smart accessories (modems), voice activation, battery eliminators, antennas, and many others.

Hands Free Speakerphone

Hands-free car kits typically include a microphone and speaker to allow the subscriber to talk to the phone without using the handset. The speaker is usually located in the cradle assembly while the microphone is typically installed in a remote microphone, usually located near the visor. Some hands-free systems have advanced echo-canceling technology to minimize or eliminate the effects of delayed echo that can be introduced in digital systems. Figure 4.5 shows a hands-free speakerphone assembly.

Smart Accessories

CDMA mobile phones have the capability to connect to various types of devices such as computer modems. When these smart accessory devices are used, they typically require an audio connection for the modem data transfer. CDMA telephones also include an all-digital data service that does not require audio channels and connects directly to a computing device without the need for a modem.

Figure 4.5, Hands-Free Speakerphone
Source: Cellport

Another optional feature includes voice activation, which allows calls to be dialed and controlled by voice commands. It is recommended that calls should not be dialed while driving because of safety concerns, but a call can be dialed (initiated) via voice activation without significant distraction.

Two types of speech recognition exist—speaker dependent and speaker independent. Speaker dependent requires the user to store his voice command to be associated with a particular command. These recorded commands are used to match words spoken during operation. Speaker independent allows multiple users to control the phone without the recording of a particular voice. To prevent accidental operation of the Mobile phone by words in normal conversation, key words such as "phone start" are used to indicate that a voice command will follow [1].

Software Download Transfer Equipment

Some mobile phones have the capability of having their operating system memory reprogrammed in the field to allow new or upgraded software to add feature enhancements or to correct software errors. The new operating software is typically downloaded using a service accessory that normally contains an adapter box connected to a portable computer and a software disk. The new software is transferred from the computer through the adapter box to the mobile phone. Optionally, an adapter box can contain a memory chip with the new software that eliminates the need for the portable computer. Changes can be made easily in the field without opening up a mobile telephone. Figure 4.6 shows how software downloading can occur from a personal computer to a mobile telephone.

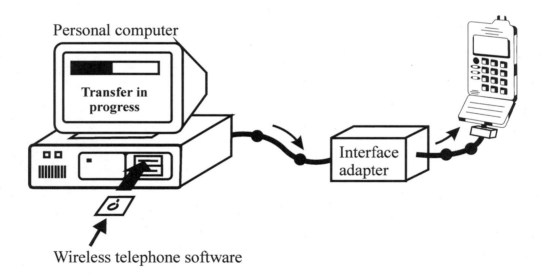

Figure 4.6, Software Downloading

Antennas

Antennas on portable mobile phones are typically integrated into the case of the mobile phone. Some phones include a coaxial antenna connection that allows the use of external antennas that may be mounted on a car.

There are several factors that will affect the performance of an antenna. Antennas convert radio signal energy to and from electromagnetic energy for transmission between the mobile phone and base station. Because there is only a fixed amount of energy available for conversion, antennas can only improve their performance by focusing energy in a particular direction, which reduces energy transmitted and received in other directions. The amount of gain is specified relative to a unity (omnidirectional) gain antenna. Car mounted antennas typically use 3 dB or 5 dB gain antennas. Portable antennas commonly use 0 to 1 dB gain antennas because people may turn the phone different angles or leave the phone laying flat on the table. Figure 4.7 shows the different types of antenna gain.

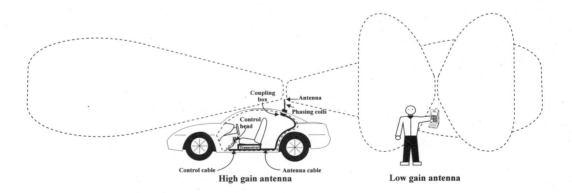

Figure 4.7, Antenna Gain

CDMA Telephone

The complexity of a CDMA mobile phone is greater than an analog mobile phone owing to the more advanced signal processing that is required. CDMA mobile phones use a 1.25 MHz wide radio channel.

Figure 4.8 shows a detailed block diagram for a CDMA mobile telephone. In this diagram, the transmitter audio section samples sound pressure from the mobile phone's microphone into a 64 kb/s digital signal. The digital signal is then divided into 20 msec groups that are sent to the speech coder for analysis and compression. The speech coder compresses the 64 kb/s signal to a data rate of 8 kb/s or 13 kb/s. Data compression rate varies with the speech activity. The lower the speech activity (e.g. silence period), the lower the data rate from the speech coder (e.g. 2 kb/s compared to 8 kb/s). The channel coder then adds error protection to some of the data bits using convolutional coding. This increases the data bit rate to 28.8 kb/s. A

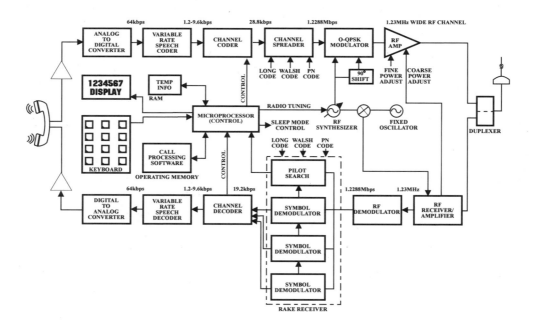

Figure 4.8, CDMA Mobile Phone Detailed Block Diagram

channel coder then mixes the necessary control information and guard and control bits. The digital signal is grouped into slots and the clock signal is increased to the radio channel data transfer rate. The burst signal is supplied to the modulator that mixes the digital with the output of the RF synthesizer to produce a 1.25 MHz wide radio channel at the desired frequency. The RF amplifier boosts the signal for transmission. When transmitting, the RF amplifier power gain is continuously adjusted by the microprocessor control section in inverse proportion to the received radio signal strength.

A received CDMA radio signal is down-converted and amplified by the RF receiver and amplifier section. Because the incoming radio signal is related to the transmitted signal, a frequency synthesizer (variable frequency) produces the fixed frequency for down-conversion. The down-conversion mixer produces a first Intermediate Frequency (IF) signal which is either digitized and supplied to a demodulator or down-converted by a 2nd IF mixer to reduce the frequency even lower. A rake receiver is used to decode delayed signals and combined them with the desired signal. This increases the quality of the received signal.

The channel decoder then extracts the data and control information and supplies the control information to the microprocessor and the speech data to the speech decoder. The speech decoder converts the data slots into 64 kb/s PCM signal that is then converted back to its original analog (audio) form.

The microprocessor section controls the overall operation of the mobile phone. It receives commands from a keypad (or other control device), provides status indication to the display (or other alert device), receives, processes, and transmits control commands to various functional assemblies in the mobile phone.

References:

1. U.S. Patent 4,827,520, Voice Actuated Control System for Use in a Vehicle, Mark Zeinstra, 1989.

Chapter 5

CDMA Networks

CDMA networks utilize a radio interface and an intelligent network to provide advanced, wireless services to a large subscriber base. This chapter describes the basic parts of the network and how they interact with each other. This chapter also describes how these networks are connected to other networks and the public switched telephone network (PSTN).

These networks are composed of mobiles, cell sites, mobile switching center (MSC), and the communications links that connect them. Figure 5.1 depicts these network components and their interconnections. The mobiles, operated by system subscribers, are the clients in the network. They communicate with cell sites through a radio connection. The network's many cell sites distribute service over the targeted coverage area. The MSC is responsible for connecting calls to other mobiles in the network, or other destinations in other networks or on the PSTN.

Cell Site

The cell site is the combination of radio transmission equipment, controllers, power supplies and backup power sources, antennas, communications structure, transmission lines, and communications links. The combination of electronics communications is called a base station.

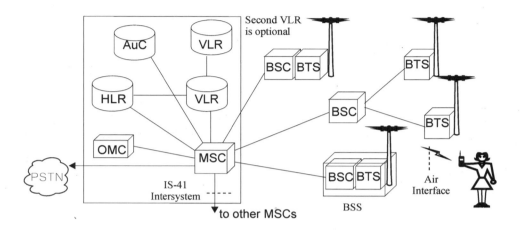

Figure 5.1 CDMA System

Base Station

IS-95 does not provide a standard for which base stations are designed, therefore a general description of the basic base station components and their functions will be provided. The base station typically consists of transmitters, receivers, controllers, and a communications interface. These devices are typically housed in a small building or weather-proof enclosure.

Figure 5.2 is a block diagram of a CDMA base station. This diagram shows that the base station can be divided into base station transceiver (BTS) and base station controller (BSC) sections. The base station transceivers convert the audio signals from the system into RF signals that are communicated to the mobile radios. The BSC is the section that coordinates multiple BTS. It is possible for the BSC to be located in a different area than the BTS units. The combination of the BTS and BSC functions are commonly called the base station (BS).

The base station transmitter converts forward link call information into Radio Frequency (RF), ready for transmission to the mobile. It contains audio processors, modulators, and power amplifiers. The audio processors convert the forward link call audio into digital information. They can also inject forward link overhead messages for the mobile. The modulator encodes this digital call information into phase shifts of the carrier frequency. The power amplifier increases the power of the signal so that it can be transmitted over the cell site's coverage area. The amplifier power will be adjusted continuously during a call to minimize excess interference.

The base station receiver converts the received RF signals into the reverse link call information. It contains power amplifiers, demodulators, and audio processors. The power amplifier increases the received, low level signals. The demodulator transforms the phase shifts of the received signal back into digital call information. The audio processors then convert the digital information into audio signals that are ready to be sent to the switching center. They can also extract reverse link overhead messages for the network.

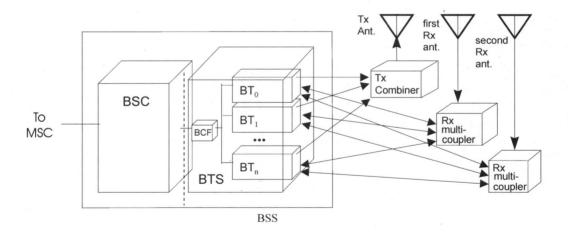

Figure 5.2, CDMA Base Station Block Diagram

Power Supplies and Backup Power Sources

Power supplies convert commercial power to AC and DC power that the base station devices can use. Backup batteries provide power to the base station when primary power is interrupted. Generators are also used when the primary power source is interrupted for a longer amount of time.

Cell Site Antennas

The purpose of the antenna system is to create the RF path between the base station and the mobile. They do this by converting electrical signals to and from electromagnetic waves. The two types of antennas that are typically used on cell sites include omni-directional and directional antennas. Figure 5.3 depicts the typical antenna that is used in cellular and PCS systems.

Figure 5.3, Cell Site Antennas
Source: SCALA

Omni-directional antennas can send or receive signals equally in all directions and can be used to create a circular coverage area. Directional antennas can send or receive signals in one direction and can be used to create a wedge shaped coverage area. Figure 5.4 shows an example of an antenna pattern for each of these antennas. Many of these wedge shaped coverage areas can be used on one site to cover in all directions

When many of these directional antennas are used on a site, it is referred to as a sectored site. Many of the wedge shaped coverage areas of the directional antennas are used with each other to cover a circular area around the cell site.

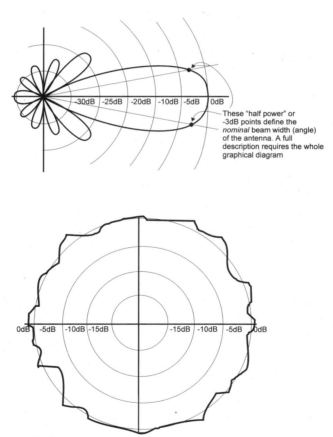

Figure 5.4, Antenna Patterns

These sectored sites focus what was covered in a circular area into a set of wedge shaped coverage areas. This helps to increase capacity by isolating the transmitted signals from one sector from the other sectors.

Typically two receive antennas, that have been separated by some distance are used on sites. This redundancy of the receive antennas is referred to as diversity receiving. Signal fading can be combated by having the ability to compare and sum the signals that were independently collected by the two receive antennas.

Antenna Placement

Cell site antennas are mounted high above the ground to increase the cell site's coverage area. They are typically mounted on a communications. Figure 5.5 depicts the different types of communications structures. Figure A is a freestanding single pole called a monopole. Figure B is a freestanding lattice tower. Figure C is a guyed

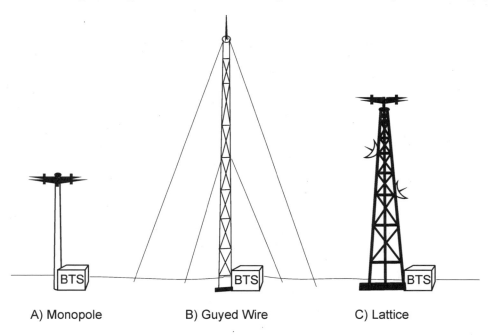

A) Monopole   B) Guyed Wire   C) Lattice

Figure 5.5, Antenna Communications Structures

tower, which is supported by cables anchored into the ground. Antennas can also be mounted to existing structures like a building rooftop or on a water tower.

Transmission Lines

Coaxial transmission lines are used to link the base station to the antennas. Both transmitted and received signals travel on these lines and experience some loss. These lines can vary in size depending on their length and the maximum signal loss limits of the specific site.

Communication Links

Communications links are set up to carry the call and control information between the cell sites and the MSC. These links can include copper wire and microwave links. A standard communication link is the T1 for the Americas and E1 for the rest of the world. This type of link can simultaneously provide multiple communications channels. Copper wire communication links are usually provided by a public telecommunications service provider that is located near the cell site's location. However, with telecommunications deregulation, it may be possible to obtain copper communications links from competitive local exchange providers (CLECs) or cable telephone companies.

A microwave link can also be used to connect communication devices together. Microwave links are wireless point to point communication devices that can also simultaneously provide multiple communication channels. The ability to use microwave transmission links depends on precise alignment of a sending antenna on the cell site with a remote antenna attached to the network. Terrain and environmental conditions can make this precise alignment extremely difficult or impossible.

The number of communication links between BSC and BTS is dependent upon how many calls can be transferred on the communications link and how many calls can be handled at the BTS. If the BTS capacity is 25 channels per sector and there are 3 sectors (total of 75) then the connection should handle at least 75 channels. The communications link connection for some manufacturers is channelized and others use un-channelized links. A channelized connection can handle the number of channels (24 for T1 and 30 for E1 less control functions). When un-channelized connections are used, it may be possible to provide for 50 to 180 over a single communication link, depending on the design.

If a BTS can handle up to 75 calls and the connection can handle 150 calls (including/excluding control and loging abilities) then a connection may be fractional. This allows two BTS's to share one connection.

Control links for the controller connections are usually on a separate link such as a dedicated leased line. This dedicated control link is important for the event of a BTS failure. If the BTS equipment fails, a control connection is available for rebooting via the separate link. This separate link can also be used for running tests without having anybody at the site to initiate the test. When new parameters and/or code needs to be downloaded from the network to the BTS, the control link can be used to transfer is needed parameters or software updates.

For call logging purposes (call detail records), the voice communications link can be used and not the control link. However, this affects capacity of the connection and call processing speed. The speed is affected due the processor capabilities at the BTS and BSC or MSC. This is determined by how fast the processor can separate logged data from voice information.

Timing Reference

Timing is extremely important in any digital system, especially for the IS-95 CDMA system with its advanced features. Different manufacturers have their own timing source that is the same for all sites. The PN codes used in CDMA are time synchronized to the start of GPS time (midnight Jan 6, 1980). A message that is continuously sent on the sync channel that includes system time.

Wireless Local Loop

Some base station equipment can be directly connected to an office telephone system or home telephones. If a home system is used, external antennas may be used to improve signal quality and increase capacity.

The box that connects the telephone system or telephone devices may be capable of handling voice, analog modems, digital modems, fax, etc. The capabilities of the wireless local loop interface are also going to be tied to what the infrastructure (network) can provide (e.g. data calls). Figure 5.6 shows a device that allows multiple telephone devices (modems, fax, phones, etc) to be connected to the cellular or PCS CDMA system. These devices should not exceed the 5 Ring Equivalent Number (REN). Five REN is what the standard phone lines in the USA provide. REN is

Figure 5.6, Dialtone Interface (QCT-6000)
Source: Qualcomm

essentially how much energy (power) is available by the telephone company for each telephone's ringer. Each phone has a REN number associated with it. Typical REN numbers range from 0.2 (for newer digital phones) to 1.0 REN for older analog phones.

Equipment Redundancy

Equipment redundancy is the inclusion of additional equipment that provides for backup operation in the event an equipment assembly experiences a failure. Redundancy can be in the following areas: BTS (transceiver) controller, timing, and cell site modems. Because there are often several BTS units installed in each cell site, this offers built in redundancy provided that the BTS units can be isolated and audio paths can be switched to other BTS units.

The controller card usually has a redundant or backup system. This backup may be an onsite redundant card that can be placed into service when the main controller fails or it can take over immediately when the main controller fails. Both controllers may have the coded loaded into them making a reboot simple and quick.

Cell site modem backup can be in the form of a spare card with modems on them or spare modems on a card that is being used. Some manufacturers do not have large capacity systems; therefore, redundancy may not be possible in which case hardware capacity is limited until such time the part can be replaced.

A timing backup system is necessary. Because timing of the CDMA base station is critical to its operation, if the timing source is lost the hardware may be designed to run for a short period of time (drift) before timing would be too far out. A timing backup card may be used. If GPS is used at the cell site then a redundant GPS may be implemented.

Cell Site Installation

The first step in cell site installation is site acquisition. After the ideal locations are found, the land is purchased or leased and all permits are obtained. Next, a power supply and backup supply is selected. This is typically obtained from another source than the system equipment manufacture. The building and towers are than installed. This is followed by the hardware installation of radio equipment and controllers. The RF power levels are then adjusted and calibrated for the specific settings given the situation where it is being installed. Over the air tests and radio covers tests are then performed. Finally, the control link is connected and the base station control software is downloading from OA&M system in the network.

Cell Site Maintenance

Once the network has been deployed, it must be maintained as cell site equipment periodically fails. Cell site alarms and test tools are developed by equipment vendors to aid in maintaining the system. Cell site visits are sometimes required to replace parts in the base station or to fix any damage to the site.

Cell site alarms are configured to alert the MSC operations center of any malfunctions in the cell site. Any problems detected by the site are reported back to the MSC for further analysis. Tests can usually be performed at the MSC to test the different functions of the base station to isolate what the problem is and to devise a solution. In the event that a problem can not be determined, a technician will need to visit the site and analyze the problem from there. Having a supply of replacement parts for your cell sites makes fixing problems both easier and faster.

Because the numerous parameters for a BTS and CDMA configurations may change, the ability to download new configurations and upgrade software via the control link is important. This is why separate contol and voice data links are important. When a network is being brought on line for the first time, the configuration is downloaded from the MSC where the OA&M (Operation, Administration, & Maintainance) center is located. The ability to download new base station control software and parameters is extremely important when the network is being expanded and new Neighbor lists and PN offsets need to be changed.

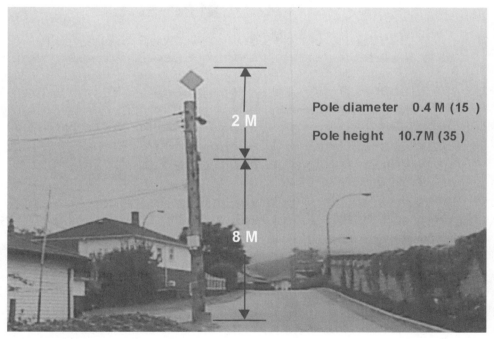

Figure 5.7, CDMA Repeater
Source: Repeater Technologies

Repeaters

Repeaters can be used to extend or redirect radio coverage. Repeaters can be a simple remote amplifier or an intelligent device. Repeaters are located within the radio coverage area of another base station. The device will amplify any incoming signal and retransmit the signal. The advantage is that the antennas can be located a longer distance from the BTS and it can also provide a greater link budget advantage. This translates into greater coverage - fewer cell sites required. A portion of the RF energy is received, processed and retransmitted. Figure 5.7 shows a CDMA repeater.

Figure 5.8 shows a block diagram of an intelligent CDMA repeater. In this diagram a receiver samples the radio signal from an adjacent base station (or another repeater). The intelligent repeater actually decodes the incoming signal, translates control commands, and re-modulates the signals on another carrier frequency.

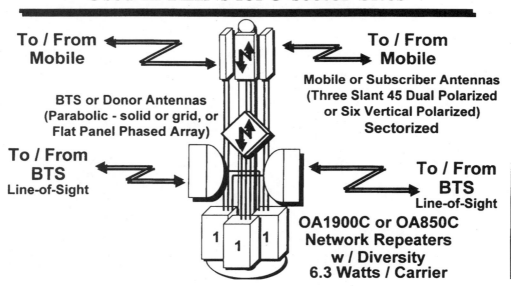

Figure 5.8, Intelligent CDMA Repeater Block Diagram
Source: Repeater Technologiesf

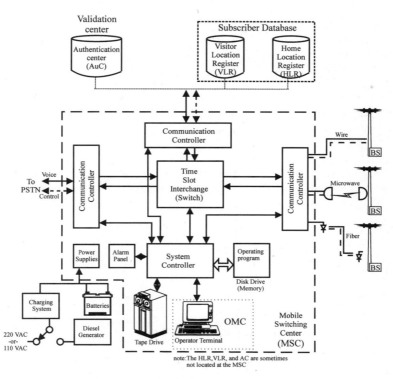

Figure 5.9, MSC block diagram

Mobile Switching Center

The Mobile Switching Center (MSC) is responsible for the communications paths and overall operation of the network. It processes calls from both land and mobile originated calls, and routes them to the proper destination. The IS-95 standard does not specify how the MSC operates and what functions it performs. The specific operations of an MSC are usually unique to the specific system equipment manufacturer. Figure 5.9 depicts an MSC's typical block diagram. An MSC contains a switching assembly, controllers, communications links, subscriber databases, operator terminals, data archive, power supplies and backup energy sources.

Switching Assembly

The switching assembly creates connections between the cell sites and the PSTN, to exchange call information. This switching system dynamically creates connections for cell site voice channels and PSTN voice channels. Figure 5.10 shows a typical switching system. In this example, multiple communication channels are supplied to a switching system in a specific sequence of time slots. Each input time slot location is received from specific communication lines (e.g. from a specific cell site radio channel). Each time slot is assigned to a temporary storage location. The output of the switching system is also sequenced into time slots. Each time slot is routed to specific connection lines (e.g. to a specific telephone line communications channel). The switching system assigns an incoming slot location to the outbound slot location via the temporary memory location.

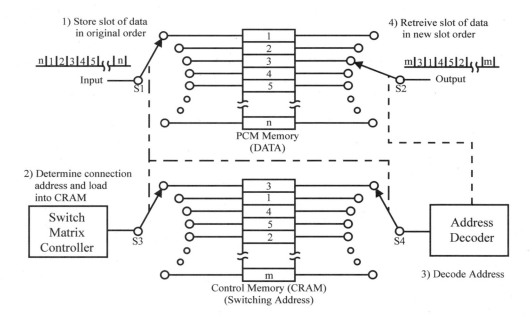

Note: Switches (S1-S4) are electronic and do not use mechanical parts.

Figure 5.10, Switching System

Controllers

Controllers perform signal routing and message processing. Controllers insert control channel signaling messages, set up voice channels, and operate the radio location/scanning receiver. In addition, controllers monitor equipment status and report operational and failure status to the MSC. Typically, there are three types of controllers: base station controller, base station communications controller, and transceiver communications controller.

The base station controller coordinates the operation of all base station equipment on commands received from the MSC. The base station communications controller buffers and rates-adapt voice and data communications from the MSC. The transceiver communications controller converts digital voice information (PCM voice channels) from the communications line to RF for radio transmission and routes signals to the wireless telephones. The transceiver controller section also commands insertion and extraction of voice information and digital signaling messages to and from the radio channel.

Billing Transfer Devices

Traditionally, tape systems have been used to transfer billing records. However, there is a movement towards electronic data interchange (EDI) to settle billing records.

Communications Links

Communications links provide a path for voice and control data to be exchanged between the MSC and the network of base stations. These communications links are typically T1 or E1 lines provided by a locally serving telecommunications company. For redundancy, these communications links can be routed a number of ways. This protects the network from going off line when an interruption occurs at some point on the communications link.

Subscriber Databases

Subscriber databases are used to identify wireless service subscribers to the network. Several of these databases are maintained and linked to the MSC. These databases include the Home Location Register (HLR) and the Visitor Location Register (VLR).

The HLR is a database that identifies all local market subscribers to the MSC. It contains information like the mobile's identification and serial numbers that are unique to each mobile phone. It also includes the subscriber's service profile. The service profile includes information about the subscriber's service options. The MSC uses this database information for subscriber authentication and call processing.

The VLR temporarily holds information about subscribers that are operating in its system. The common misunderstanding about the VLR is that it only holds information about customers that are visiting its system. A VLR holds all of the customers that are operating in its system including home customers. The VLR keeps track of the latest movements of all the customers operating in the system. To keep the amount of information that is held in the VLR to a minimum (more memory requirements requires more processing power and hard disk storage), the VLR is periodically refreshed (usually daily).

Operator Terminals

The operator terminals at the MSC are used for control and maintenance of the network. Maintenance personnel can use operator terminals to monitor alarm signals, test cell sites, and modify databases and switching settings.

Backup Energy Sources

Backup energy sources are necessary to maintain network service in the event of commercial power interruption. Batteries are kept charged up while power is available, and are used to power the MSC during brief power outages. For longer outages, generators are installed to power the MSC for extended power outages.

Data Archive

A backup data archive is maintained so that the MSC can quickly recover from any outages. All information is backed up onto tape drives and can be retrieved at any time.

Network Redundancy

Because the network system is the heart of the wireless system, various parts of the network must have automatic equipment backup. Perhaps the most important equipment for redundancy is the HLR. Most systems include a second HLR that holds the same information. As new information is entered or updated in the HLR, both the main and backup HLR are simultaneously updated. In the event of a HLR failure, the alternate HLR will be automatically brought on line when the main HLR fails.

Redundant communications lines are sometimes used to increase network reliability. However, up to 20-30% of the network cost can be leased lines [1].

Billing records are also of critical importance because wireless carriers want to get paid for providing service. Billing storage systems typically have some form of high availability and redundancy.

Network Planning

CDMA network design is a very intricate process. The cell site coverage areas are first designed. The individual cell sites must be positioned and configured so that they do not overlap too much while leaving no coverage gaps in the desired coverage objective. Antennas and transmission lines must be selected and positioned to perform adequately in the cell site coverage area's environment.

Cell site placement, signal propagation, traffic planning, system testing and optimization, and future expansion plans all play a role in system design. Signal propagation varies with terrain, morphology, and even the change of seasons and has a dramatic affect on a cell's coverage area. Traffic loading must be considered when positioning cell sites to ensure that the system capacity will not be exceeded to quickly. System testing and optimization will perfect the network by maximizing capacity and minimizing the occurrence of coverage holes.

Cell Site Placement

Cell site placement is a very complicated part of the network design process. Cell sites must be designed and selected carefully so that they conform to FAA and FCC guidelines along with local zoning restrictions. Cell site radio coverage areas must overlap the system must be designed so that cell sites work with each other (hand-off correctly) to cover the entire service area. To assist in the cell site placement process, radio propagation drive testing for cell site candidates and the use of network radio signal planning software can help.

The first step in the design process is to plan a theoretical cell site design, by geographically placing cell sites over the coverage objective area. Network planning software can be used to help with this stage of the design. This software can simulate the coverage properties of the cell sites. The next step is to find actual places near each of the theoretical cell sites to build real sites. This can be somewhat challenging because some of the theoretical sites may be located in areas where no sites can be constructed like residential areas. This is why the initial cell site design is so important.

Reviewing local zoning ordinances can help to make site candidate identification a much easier process. Once cell site candidates have been selected, the most suitable candidate is identified for further development. A drive test can be performed on the candidates to determine how the site will perform in the network. A crane can be used to raise a test transmitter to the height of the proposed site, and a vehicle can drive in the area to test the signal propagation properties of the site. This data is used to tune radio propagation models in the network planning software, to more accurately simulate the cell site's coverage. FAA and FCC guidelines must also be followed when designing the network. Some FAA rules can limit communication structure height in certain areas. FCC rules regulate transmitter power levels and signal exposure levels.

Signal Propagation

Radio signal propagation must be analyzed for each cell site to determine its coverage characteristics. Path loss and fading affect a radio signal's propagation characteristics. Path loss reveals the rate at which a radio signal weakens as it travels away from its source. The fading characteristics of a radio signal account for irregular signal changes over short distances.

Path Loss

Path loss describes the energy losses that a radio signal encounters during propagation between the transmitting and receiving devices. Many different factors have an effect on path loss. This includes terrain type, foliage density, and environmental clutter. Flat terrain allows signals to propagate freely without running into the ground and provide more even coverage. Signals can pass over valleys or be blocked by hills, which can create holes in system coverage. Dense foliage absorbs radio signals and significantly diminishes their strength. Environmental clutter refers to variations in the environment which include urban areas densely filled with large buildings, suburban areas scattered with shops and houses, and rural areas containing some spaced out houses and highways.

Path loss models can be created to accurately predict how signals will propagate in each cell site's coverage area. They are usually generated with a software design tool and tuned with actual test data collected from sites. These path loss models are then used to make predictions of how the network will operate and can help forecast any coverage or interference problems.

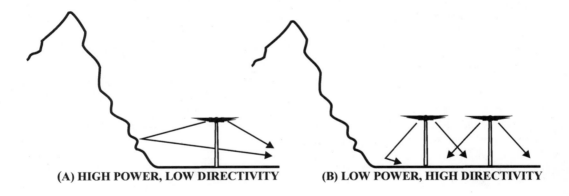

Figure 5.11, Multipath Reflections

Fading Characteristics

Radio signal fading is caused when a radio signal interacts with a delayed version of the radio signal. The delayed signal can be the result of a signal reflecting on the ground or some environmental clutter. This type of fading is called multipath fading or Rayleigh fading and results in very quick fades that change with movement. Figure 5.11 illustrates a signal directly propagating from cell site to a mobile and a multipath reflection.

CDMA networks can overcome some of the effects of Rayleigh fading by using a rake receiver. The rake receiver allows the receiver to accept multiple signals, whether they are from different sectors or many delayed signals from the same sector. This receiver can correlate these different signals and combine them.

System Capacity

The overall capacity of a wireless system depends on two key factors. These are spectral efficiency and interference levels. Spectral efficiency refers to how many simultaneous conversations that can occur, in the same frequency band, at the same time. The other factor is the presence of interference in the same frequency band (co-channel interference).

In CDMA systems, all of the conversations occur in the same frequency band. This means that each cell site transmits on the same frequency and can become destructive interference to other adjacent cells. Thermal noise, the noise caused by increased particle movement in the environment due to heat energy, also contributes to interference. The quality of a given traffic channel signal can be evaluated by examining the ratio of the energy in each received CDMA chip to the present interference level, which is written as:

Ec/Io

The Ec/Io ratio is expressed in decibels rather then using a numeric ratio. This is a very important parameter in CDMA. It is used to direct handoffs between cell sites by evaluating the signal strength received from adjacent cells.

Traffic Planning

Before designing the network, subscriber traffic loading must first be projected and analyzed. This is to ensure that the network that is built will be able of providing service to all its subscribers. Census information, wireless service market penetration, and other relevant data are analyzed and subscriber counts are formulated for the different parts of the service coverage area. Locations like downtown areas of cities and business parks are typically projected with a higher subscriber count than residential and highway areas. The network design takes these projections into account when planning the density of cell sites in these areas.

Strategic Planning

Strategic planning for a cellular service provider involves setting company goals, such as subscriber growth, quality of service, and cost objectives. It also involves making plans for obtaining those goals. Building and expanding a cellular system requires collecting demographic information, targeting key high traffic locations, selecting potential cell and MSC sites, conforming to government regulations, purchasing equipment, construction, and testing validation.

Most providers gather physical and demographic information first. For example, transportation thoroughfares, industrial parks, convention centers, railway centers, and airports may be identified as possible high usage areas. Estimates of traffic patterns are used to help target coverage areas for major roadway corridors. Terrain maps, marketing data, and demographic data are all used to divide the cellular system into RF coverage areas. The object is to target gross areas where cell site towers may be located. The raw data needed might include system specifications, road maps, population density distribution maps, significant urban center locations, marketing demographic data, elevation data, and PSTN and switch center locations.

For the US market, government regulations include quality of service (typical limiting the blocked call attempt ratio to 2%, or P02 grade of service) and time intervals for service offerings [2]. While business considerations may indicate that radio coverage is not necessary (e.g. an unpopulated rural area), government regulations may require that area to be covered within a specified period.

After systems are planned, equipment manufacturers and their systems are reviewed and purchase contracts are signed. During various stages of equipment installation, validation testing is performed to ensure that all of the planning goals are being realized.

After the system is planned and cell site locations are selected, RF simulation begins. Calculations based on antenna elevation and terrain data are used to estimate expected signal strengths and quality levels.

The results of such calculations are rendered graphically onto transparent overlays that can be placed over standard topographical maps (published by the US Coast and Geodetic Survey in the US, and by similar government agencies in other countries). Typically, different colors used on these overlays indicate different values of signal strength levels. System simulations may predict estimated signal coverage and performance levels, but to be certain, temporary cell sites are often tested using a crane to lift a temporary antenna to the planned tower height. Theoretical calculations are often imprecise everywhere by a relatively uniform dB error, which can be determined only by comparing theory and experimental measurement. Once the proper dB correction factor is known from this comparison, the theoretical calculation can be used for evaluation of other base antenna locations in the cell with considerably improved precision.

PN Planning

CDMA channels use Short PN offset similar to AMPS use of frequency plans. A PN offset is a delayed version of a cyclical, pseudo-random code. Each of these offset versions of the code appear to be unique codes. These offset values are assigned to each sector in each site. Cell sites can share the same PN offset values, but must be separated by enough distance to ensure no interference. PN offsets identify the sector to a mobile, and play an important role in handoff directions.

Neighbor List

The neighbor list is a network parameter used to direct handoff. Each sector has a neighbor list associated with it that defines what other sectors a mobile can handoff to while communicating with it. The neighbor list must be examined carefully. By omitting an adjacent sector from a neighbor list, the mobile cannot handoff to it and sees it as interference.

Soft Handoff

Soft handoff between cell sites is unique to CDMA. As a mobile telephone moves from one sector's coverage are to another, it makes a connection with the new sector before breaking the connection to the old one. With this type of handoff, the call is never broken between sector handoff. This differs from other wireless standards that use a hard handoff. With a hard handoff the sector briefly breaks the connection with the mobile before reestablishing the connection with another sector. Figure 5.12 illustrates the difference between these two types of handoffs. The CDMA technology allows a mobile to communicate with up to three different sectors at once. Both the mobile and network can benefit from this by combining the many signals that is referred to as soft handoff gain.

System Testing and Optimization

System testing and optimization is a process that measures the quality of the network and adjusts it to perform at maximum efficiency and quality. Specially

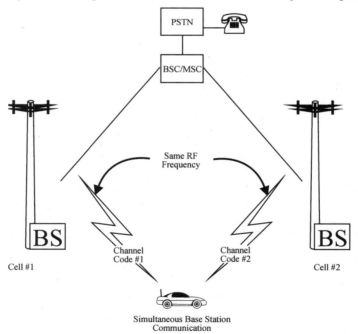

Figure 5.12, Soft Handoff compared to Hard Handoff

designed software and hardware is used for test and analysis. Vehicles are loaded with this test equipment and are driven through the coverage area to test the network quality. Data is collected on these drive tests and reviewed. The factors that are evaluated and their definitions include:

Forward Channel Signal Strength - The received power in the forward link channel.
Forward Frame Error Rate (FFER) - The percentage of the forward transmitted frames the are corrupted.
Reverse Frame Error Rate (RFER) - The percentage of the reverse transmitted frames that are corrupted.
Reverse Transmitted Power - The total power that the phone is transmitting.
Forward and Reverse FER (Frame Error Rate)
Ec/Io - The CDMA signal quality defined as the CDMA chip energy divided by the present interference.
Soft Handoff Rate - The percentage of time a mobile participates in soft handoff.
Dropped Calls - The number of times the network drops communication with the mobile.

Areas of poor network performance are identified and adjustments are implemented to fix the problems. These adjustments can include antenna changes, antenna tilting, antenna re-orientation, or adjustment to a number of system parameters. The system is then tested again to ensure that the adjustments fixed the problems, without creating any new ones. This process is repeated until the network performs to the service provider's specifications.

System Expansion

As the number of CDMA service subscribers grows, the network must also grow to accommodate their additional traffic requirements. There are several options that can be considered when expanding the capacity of an existing network. These options include cell splitting and carrier addition. Cell splitting is making multiple new cell sites to aid an existing cell handle its traffic loading. Figure 5.13 depicts how cell splitting works.

First the existing cell is altered to reduce its coverage area. One or more new cell sites are built to cover the areas that are not being covered by the altered existing cell site. The other typical solution to capacity problems is adding an additional carrier. Another frequency channel is deployed using the same cell site design. To use this method, you must own the license for additional frequency.

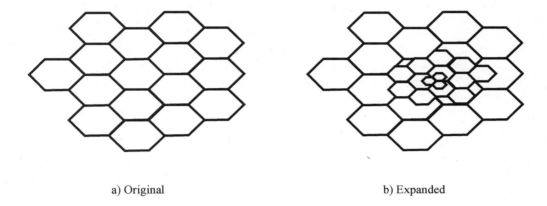

a) Original   b) Expanded

Figure 5.13, Cell Splitting

System Features

System features are usually sold as software modules. Adding system features (e.g. three-way calling) usually requires a new download of new software. Ideally, this is easily done and is always backward compatible. As systems change and unskilled operators are used, downloading and installing new features can be challenging. Carriers often desire to add unique features to better compete against their rivals. These features can be developed using Wireless Intelligent Network (WIN). WIN is the use of network software standards (they can be proprietary) that allow carriers to implement new features and capabilities as needed without purchasing software from systems manufacturers. As of 1999, there was no formal standard for WIN systems.

More and more the features and processes that had been accomplished by hardware are being replaced by software. Manufacturers are replacing custom application specific integrated circuits (ASICs) with digital signal processors (DSPs) that can have their operational program stored in random access memory (RAM). This allows for a wide variety of system features that can be easily upgraded. However, new features often require extensive testing to ensure reliable operation.

When a new feature is being developed a manufacture is wise to test it out in a lab where it can be debugged with the current hardware configurations it will be used on. The amount of testing is usually based on the number of lines of software operating code that is used. Some manufacturers do a limited amount of test while others let the carriers debug the code by running it. It can take as much as a year to test a feature dependent on the interaction of the new feature with other services.

Network Options

When deciding to design and implement a CDMA network, there are several options that are available. These options include deploying a completely new CDMA network and making a CDMA overlay network for an existing network. A new network involves the entire design process from initial site identification through optimization. An overlay system involves altering an existing network to include CDMA network equipment on existing cell sites. An example of this is creating a CDMA overlay design for an AMPS network. Multiple AMPS channels are removed from AMPS service and used for CDMA channels. The frequency plan of the network is then redesigned so that they don't interfere with the CDMA service.

Deployment Considerations

Soft Capacity

Carriers frequently experience unexpected spikes in traffic levels and operators need to be able to respond immediately to relieve congestion. Selected vocoders used in CDMA are capable of instantly maximize the capacity of their networks during a particular week, day or hour.

Political conventions, sporting events and natural disasters create traffic surges that temporarily exceed an operator's planned network performance. QUALCOMM's Global SmartRate is a vocoder feature allowing operators to temporarily boost capacity on their system by instantly altering the system wide vocoding rate. This makes it possible to balance the relationship between voice quality and capacity. When the Global SmartRate feature is used operators can fine-tune the network to allow for temporary spikes in traffic. This soft capacity feature allows the carrier to direct expansion of the network to areas where capacity increases are required on an ongoing basis. This feature offers multiple vocoding rates between 8 Kbps and 13 Kbps. The overall system capacity can be maximized to the precise level needed immediately, or for a specific duration.

QUALCOMM's Global SmartRate feature on the other hand, does not require new or upgraded handsets and can be used immediately with today's 13 Kbps-compatible CDMA handsets.

Voice quality and capacity in a CDMA network enable operators to maximize each area of performance based on their particular network requirements. Most CDMA networks today use a 13 Kbps vocoder providing near landline voice quality. The soft capacity is a trade off between quality and capacity. As the vocoder rate drops the quality of the call may slightly deteriorate while the capacity will increase.

Guard Bands

While the 25 MHz x 2 cellular band allocation on the 800 MHz band remains the same, the frequency allocation for use by CDMA systems has been divided so that 9 or 10 of the 1.23 MHz CDMA channels can be allocated in the A or B frequency bands. Because CDMA channels have more bandwidth than analog channels and require a guard band at the ends of the spectrum, the allowable channel assignments for CDMA channels are 1013 through 1023 and 1 through 777. The center frequency is used for the CDMA channel assignment. In TIA/EIA-95 cellular band there are two primary designated CDMA channels and two secondary CDMA channels. The primary and secondary CDMA channel designation ensures that a MS from one carrier can quickly and easily find the CDMA radio channel in another carriers network. Channels 283 & 384 are the primary and channels 691 & 777 are the secondary. Figure 5.14 illustrates the CDMA channel guard bands.

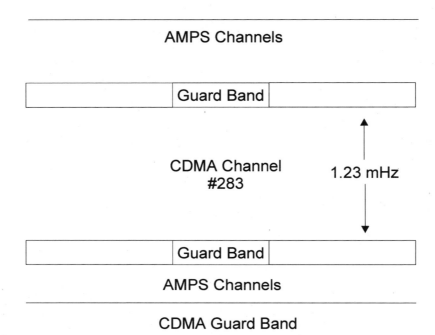

CDMA Guard Band

Figure 5.14, CDMA Channel Guard Bands

Guard Zones

TIA/EIA-95was designed to be compatible with the existing analog network. In other words CDMA fits into the same spectrum and channel designations without any changes to the spectrum or channel allocation. The carrier deploying CDMA where there is existing AMPS service requires that 42 consecutive analog radio channels be cleared, plus guard band, in each cell that will have CDMA.

The CDMA overlay on AMPS needs a guard zone between the edge of CDMA coverage and AMPS cells using the CDMA channel for AMPS service. Figure 5.15 illustrates the CDMA channel guard zones.

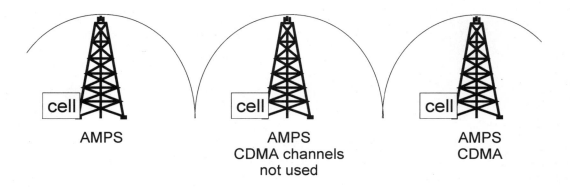

Figure 5.15, CDMA Channel Guard Zones

Network Optimization

Following a deployment the network must be fine-tuned for maximum capacity and performance. The network performance must be optimized in balancing the forward and reverse link. This will ensure consistent coverage and minimal failed access attempts.

References:

1. Personal interview, Industry Expert, 30 Dec 1998.
2. FCC Regulations, Part 22, Subpart K, "Domestic Public Cellular Radio Telecommunications Service," 22.903, (June 1981).

Chapter 6

CDMA Testing

Due to the large number of mobile phone model types in the marketplace today, testing the performance and reliability of mobile phones has become an integral part of the manufacturer's job. Testing focuses on verifying the operation and different tolerances, associated with mobile phones and their system equipment, are up to the correct standards. Analog and digital signals are test measured for different parameters, but the resulting performance characteristics (for example, clarity, robustness, etc.) are similarly affected for both signals by the different parameters.

Analog -vs.- Digital Testing

The testing procedures and the required test equipment differs between CDMA (digital) radio testing and analog radio testing. Because IS-95 CDMA can combine digital and analog functions (dual mode), different tests must be performed fore each mode of operation if the analog (AMPS) function is included (some mobile phones are CDMA only).

Fortunately, many manufacturers have produced a single piece of test equipment that capable of testing both modes of the phone. However, the test equipment will often use completely separate subsystems for each type of test. Measurements for digital systems are taken by digitizing a portion of the signal and using digital signal processing to extirpate information within the given parameters. Measurements of the analog mode parameters can be attained by using traditional audio and RF measurements.

To simplify the test design and eliminate the need to have a standardized test interface connection, a loopback mode is designed into CDMA radios. This allows the mobile phone to report back its performance to the test equipment via the radio link.

Transmitter Measurements

Transmitter measurements ensure a mobile phone correctly transmits within its prescribed bandwidth, has power control of its RF output power level and if the mobile phone can successfully inhibit its transmitter in the event of radio interference or equipment failure. RF power, frequency accuracy, spurious emissions and modulation quality measurements are the components typically involved with transmitter tests. The complexity of the CDMA radio system additionally involves testing for CDMA handoff, coding accuracy, range and response time of power control, access probe output power testing and minimum controlled and standby RF power level. These tests will determine if the mobile phone can correctly operate within the bandwidth, frequency and RF Power constraints placed on it by the industry's operating standards.

Transmitter RF Frequency and Phase

The frequency of digital mobile phone tends to be more complicated to measure as compared to analog phones. To measure the frequency of an analog phone, the transmitter is enabled without any input audio signal and a frequency counter is used to average the frequency over an approximately 1 second. This process is different when measuring the frequency and modulation accuracy of the CDMA radio transmitter.

In a CDMA mobile phone, the transmitter frequency is locked to the frequency of the incoming (received) signal. A stable forward channel frequency must be provided to mobile phone prior to measuring the frequency and phase accuracy of a CDMA mobile phone. The received code must be decoded and compared to the input reference frequency to determine its accuracy.

RF Power Measurements

Analog radio channel power measurements can be performed on a standard power meter. A standard power meter uses a crystal diode to detect the voltage across a known resistive load. This voltage is supplied to a meter that is calibrated to indicate the power that is represented by the sensed voltage.

Because CDMA mobile phones have the ability to transmit in bursts dependent on the data transfer rate, a standard power meter should not be used to measure the RF power on a CDMA radio channel. This is because the pulsed transmissions of the CDMA mobile phone require a different type of power meter. Standard power meters that use a crystal detector only sense peak power. High-speed power analyzers are the preferred way to test the power of a CDMA mobile phone. These specialized instruments use certain power meter sensors to gauge the immediate power level and to plot it on an oscilloscope type display.

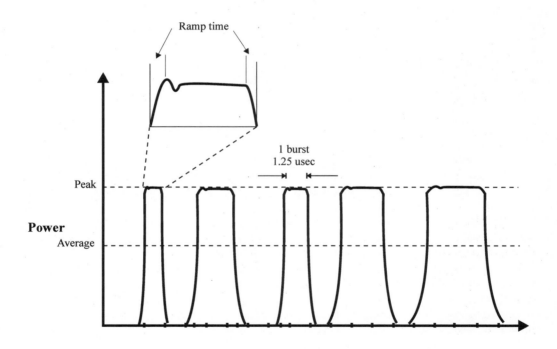

Figure 6.1, CDMA RF Power Measurement

These high-speed power analyzers can then be used to plot the RF power over the burst periods burst (1.25 usec) or a thermal power meter can be used to measure the real RF power level. However, these meters tend to be expensive. Many newer single box test systems measure the burst power by digitizing an entire RF signal burst and using digital signal processing to calculate the power level. Figure 6.1 shows a sample RF power measurement. This diagram shows that when a CDMA mobile transmitter is communicating at less than full rate, the transmitter output occurs in burst of 1.25 usec increments. A traditional power meter will measure the peak power level of the transmitted signal.

A mobile phone is tested to ensure its dynamic range of open loop output power control and ability to change quickly (time response) is sufficient. The amount of open loop power gain is determined by the signal strength level of the received signal. This is an inverse relationship where the lower the received signal strength level, the higher the output power of the mobile transmitter.

To test open loop RF power gain, this involves applying reference RF signal level to the receiver of the mobile phone. The reference RF signal level is then set to various levels and the RF transmitter output power level is measured. In addition to the RF power open loop gain measurement, the amount of time it takes for the RF power level to change after a change in the input RF signal level is also measured. Because the controlled power level adds or subtracts from the open loop power level gain, the power control bits are alternated (0 and 1) to inhibit any changes due to the control power level gain.

The closed loop power control is verified to ensure the system can adjust (fine tune) the power level of the mobile phone. The closed loop power control function is tested by providing a sequence of 100 consecutive power up (0) and power down (1) power control bits and then monitoring the RF output of the mobile phone.

The RF power control system is also tested during the system access service request (access probes). To gain access to the system, the mobile phone begins transmitting at pre-determined power levels and in a sequence of service request probes. These RF transmission probes gradually increase their power level until the system responds to the service request or a maximum limit is reached. A base station simulator provides the mobile phone with the access probe parameters (e.g. initial transmit power level) and then test measures the number of access probes and power levels of the access probes initiated by the mobile phone during access attempts.

The maximum RF output power of the mobile phone is verified by first supplying a low level RF input signal to the mobile phone (-104 dBm) and then providing a continuous sequence of power control bits (0). This forces the transmitter power gain to its maximum level. The RF power is then measured.

Modulation Quality and Frequency Accuracy

Information is transferred via a RF carrier via a form of modulation (e.g. frequency or phase). Because the quality of a modulated signal can vary, testing needs to be performed so that a standard modulation quality is met. Different tests are used in analog and digital modes to measure the quality of the modulation. A basic FM receiver is used, in analog mode, to test the quantities of: transmit deviation in both voice and data modes, residual deviation with no audio input, SAT deviation, distortion, SINAD, compression and limiting.

The first part of measuring the modulation quality for CDMA transmission is the time reference measurement. The time reference measurement provides an indication of the amount of time error the mobile phone has when receiving multipath signals. To test the timing reference, a base station simulator provides a continuous reference timing signal that has alternating delay periods of 10 chips. A special meter compares the reference signal to the timing reference provided by the mobile phone. The next test verifies the waveform quality factor. This test determines the frequency accuracy and the transmit timing error.

Adjacent and Alternate Channel Power

Small amounts of radio energy are produced outside the designated radio channel when a radio carrier is modulated. This out of bandwidth energy can cause interference with adjacent channels or alternate channels (two channels away from the center frequency). Adjacent channel interference is when that radio energy falls in the channel that is directly above or below the designated channel. Alternate channel interference, however, is when the radio energy flows into a radio channel that is two channels above or below. Both types of interference are measured the same way. The mobile phone is set-up at full transmitter power and the radio energy is measured outside the channel bandwidth. Figure 6.2 shows the allowable bandpass power for adjacent and alternate channel power. This is called spurious emissions.

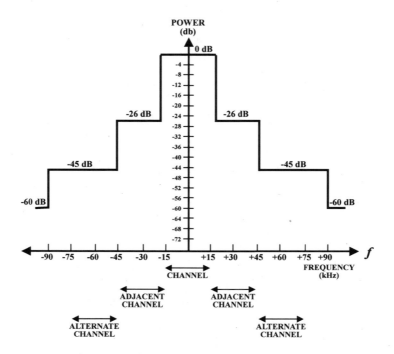

Figure 6.2, Spurious Emissions Limits

Handoff Testing

There are several types of handoff supported by the IS-95 CDMA system. Performance tests ensure that a CDMA radio is capable of CDMA to CDMA hard handoff. A CDMA hard handoff occurs when a mobile radio is assigned from an existing CDMA channel to an independent (different pilot PN offset indices) CDMA channel. This could be the result of handoff command to a new frequency or to an adjacent CDMA system.

To test for hard handoff, two base station simulators are setup with different PN offset indices. The mobile phone is setup in the loopback test mode. The serving base station sends an Extended Handoff Direction message and the amount of handoff time is measured. The time is calculated by sensing when the transmitter power drops below -61 dBm and when the mobile phone's transmitter is enabled on the new channel. This maximum time for hard handoff is 140 msec.

Coding Accuracy

The coding accuracy test measures the ability of the mobile phone to properly encode data. This test is performed by transmitting data by the base station simulator to the mobile phone and comparing the looped back information to the data that was sent to the mobile phone. This test requires no errors over the test measurement period.

Minimum Controlled and Standby RF Power Level

The mobile phone must be capable of reducing its output power so it does not interfere with other mobile radios. There are two cases for minimum transmitted power; when the unit is transmitting (called controlled RF) and when the mobile phone is in the idle mode (standby) or in the gated mode (no voice activity and the transmitter is off).

When the mobile phone is transmitting (e.g. a call is in progress), the mobile phone will dynamically reduce its power as the received signal level increases (e.g. the mobile phone moves closer to the base station). The mobile phone will also adjust its output power when the system sends power control bits (1) indicating it to decrease its transmitted power. The minimum controlled power level is tested by supplying a very high reference input signal and by continuously sending power control bits (1) indicating to the mobile radio to decrease its transmit power level. The maximum amount of controlled RF output power is -50 dBm. When the mobile phone is in idle mode (standby) or gated mode, the transmitter should be inhibited. The maximum output power when in the idle (standby) mode is -61 dBm.

Conducted and Radiated Spurious Emissions

A mobile radio produces a variety of radio signals within its chassis or housing that are not part of the desired signal to be transmitted (e.g. reference clocks and RF signals that are supplied to a mixer). Ideally, these signals should be contained within the mobile phone. If some of these RF signals may radiate out of the case through connectors, unshielded cases and other parts of the mobile phone, this is called spurious emissions. There are two types of spurious emissions: conducted and radiated.

There are two tests that are performed to ensure unwanted radio energy is not emanating from the mobile phone's transmitter. Conducted spurious emissions are measured by connecting a spectrum analyzer to the mobile phone's antenna when the transmitter is on. The spectrum analyzer is then used to scan outside the transmitted frequency band to view unwanted RF signals. Radiated spurious emissions can emanate from different parts of the mobile phone. For this test, an antenna is connected to the spectrum analyzer and it is positioned at various points around the mobile phone. The spectrum analyzer is again used to scan outside the transmitted frequency band to locate unwanted RF transmissions while the mobile phone is in the transmit state.

Receiver Measurements

Receiver tests determine if the mobile phone can correctly receive and demodulate signals within its own bandwidth and reject interfering signals. Receiver measurements for analog signals involve RF sensitivity, demodulated signal quality, co-channel signal rejection, adjacent channel signal rejection and intermodulation rejection. There are additional receiver measurement tests for the CDMA radio portion that include system acquisition (pilot detection), digital demodulation quality, open loop and closed loop power control, receiver dynamic range, single tone desensitization and radio channel supervision.

Testing a CDMA radio requires new types of equipment and measurements. To simulate other radios operating in the system, CDMA receiver tests generally require an additive white gaussian noise (AWGN) generator. This noise generator simulates the radio transmissions from nearby mobile phones that are not synchronized with the mobile phone under test.

In place of a signal to noise and distortion measurement, distortion is measured using a measure of the number of frames that are received in error when a good quality signal is present. This is called the frame error rate (FER). Some receiver tests require the use of a RF signal fading simulator to test the performance of the receiver in conditions similar to what is experienced in the field. The AWGN generator and the RF signal fading simulator are used in many of the receiver tests.

System Acquisition

One of the most basic functions of a mobile phone is to acquire the pilot of a CDMA radio channel and register with the system. The first two system acquisition tests verify if the mobile phone is capable of system acquisition during the idle mode. There are two idle mode system access tests: the idle handoff in the slotted mode and idle handoff in the unslotted (discontinuous reception) mode. Additional system acquisition tests are performed to determine if a mobile phone can acquire the pilot channels that are supplied by a neighbor set list supplied by the serving base station. Finally, the mobile phone is tested for its performance when a pilot signal is lost (typically when signal loss occurs in the handoff condition).

The first part of each test verifies that a mobile phone will successfully access a system by sensing a pilot channel, acquiring system information and registering with a base station when a pilot signal is of sufficient signal strength level. The test also ensures that a mobile phone will not constantly acquire new pilot channels unless the new pilot channel is at a sufficient signal strength difference between the two pilot signals. This prevents a mobile phone from continuously performing idle handoffs with the system. Figure 6.3 shows the basic test setup for system acquisition testing.

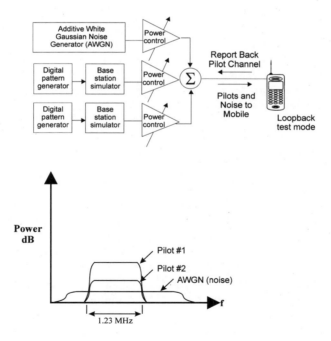

Figure 6.3, System Acquisition Testing

Power Control

The CDMA receiver must be capable of decoding the power control messages that are used to control (fine tune) the mobile phone's transmitter. This is especially true during soft handoff when multiple signals are received that may have different power control commands. The receiver power control bit test determines how the mobile phone will operate when it receives power control bits and what happens when different power control bits are received while the phone is in the soft hand-off condition (diversity combining of power control bits). The test involves sending a series of power control bits and monitoring the transmitter power level as these bits are received.

Single Tone Desensitization

The single tone desensitization test ensures that a CDMA receiver can continue to operate when a high power narrowband interfering signal is received in the 1.23 MHz CDMA channel bandwidth. To test the ability to reject a narrowband interfering signal, a known digital signal that is supplied by a base station simulator is combined with an interfering narrowband continuous wave (CW) signal. The base station simulator provides a series of messages and the mobile phone then responds back with the number of frames successfully received. The RF signal source is adjusted to the interference test levels and the FER is checked by comparing the number of frames sent to the number of frames successfully received.

Radio Channel Supervision

The mobile phone is tested for the ability to monitor the status of the paging channel when it is in the system access state. This test ensures that a mobile phone can sense (determine) if a system is busy prior to its attempting to access the system. If the mobile phone could not determine that the system was busy, it may attempt to transmit an access request at the same time as another mobile phone is communicating with the system. The test is performed by verifying that the mobile phone disables its' transmitter within a specific time period after the system has indicated it has become busy servicing another mobile phone.

Analog Receiver Sensitivity

The RF sensitivity an analog (FM) receiver is measured by supplying a FM modulated signal to the mobile phone and measuring the audio quality as the RF signal strength level is decreased. The RF sensitivity of the receiver is determined when the maximum amount of distortion is measured at the audio output of the mobile phone. This is called a signal to noise and distortion (SINAD) test. To perform a SINAD test, a RF test signal is modulated with a fixed tone (usually 1 KHz) and supplied to the mobile phone's antenna. An audio distortion meter is connected to both the audio signal source and the audio output of the mobile phone. The distortion analyzer observes the difference between the transmitted audio signal and the received audio signal. The RF signal strength is steadily lowered until the distortion level surpasses the distortion tolerance. When this occurs, the RF signal level determines the sensitivity of the receiver. Figure 6.4 shows a block diagram of the receiver sensitivity measurement test.

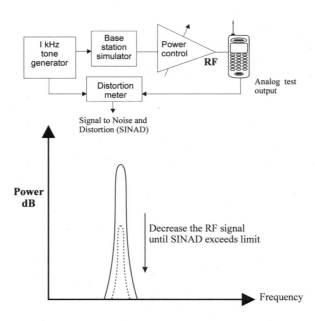

Figure 6.4, Analog Receiver Sensitivity Measurement

CDMA Receiver Sensitivity and Dynamic Range

To determine a mobile phone's RF sensitivity and its receivers' dynamic range on a CDMA radio channel, a receiver sensitivity test will measure the mobile phone's frame error rate (FER) in a variety of RF signal conditions. To test the RF sensitivity, a known digital signal (messages) is supplied to the base station simulator at a very low RF signal level. The mobile phone then responds back (loopback test mode) with the number of frames successfully received. The FER is checked by comparing the number of frames sent to the number of frames successfully received. The dynamic range of the receiver is verified by supplying other RF signal levels to the receiver. The receiver must be capable of receiving low to high levels of RF signals.

Demodulated Digital Signal Quality

The performance of the digital signal demodulation is measured through the use of loopback testing. To perform the demodulation signal quality test, a base station simulator sends a standard digital signal source (messages) that supplies known segments of information to the mobile phone. The mobile phone attempts to receive the messages and reports back the number of messages successfully received. The mobile phone must be capable of demodulation of a paging channel and traffic channel in a variety of signal interference conditions. This is why the demodulation tests include conditions with background noise (AWGN), multipath fading signals and poor quality signals (such as experienced during handoff conditions).

Co-channel Rejection

The ability of a mobile phone to differentiate between a desired radio signal and a weak interfering signal that is operating on the same frequency is called co-channel rejection. To determine the co-channel rejection ability of an analog receiver, a reference signal with a modulated audio tone (typically 1 kHz) to the receiver and combining an interfering RF signal that is operating on the same frequency. The desired output of the mobile phone is then compared to the desired audio signal to sense distortion. The RF signal level of the interfering signal is gradually raised until the audio distortion exceeds a maximum limit. This is the amount of co-channel interference the mobile phone can accept. This is similar to the receiver sensitivity test.

Figure 6.5 shows how co-channel rejection is measured for an analog receiver. This diagram shows that a communications link is established with the mobile phone.

Another communications channel is established on the same frequency using the same frequency at a very low signal level. This signal level is gradually increased until the SINAD exceeds a specific threshold. The difference between the desired RF signal level and the RF signal level of the co-channel interference signal is the co-channel rejection ratio.

Adjacent Channel Rejection

Adjacent channel rejection is the ability of a mobile phone to distinguish between its current signal and a strong RF signal on an adjacent radio channel. The measurement test of this ability is similar to the co-channel rejection measurement test.

Figure 6.6 shows how to test for adjacent channel rejection for an analog receiver. In this diagram, a reference digital signal is applied to the receiver with a known

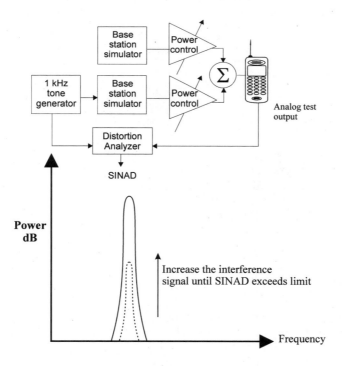

Figure 6.5, Analog Co-Channel Rejection Measurements

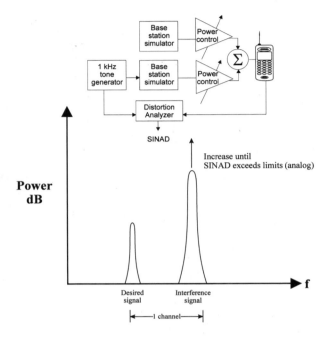

Figure 6.6, Adjacent Channel Interference Rejection Measurements

digital sequence. An interfering RF signal is applied at a radio channel above or below with a different digital pattern. The maximum tolerance level of adjacent channel interference is determined by increasing the RF signal level until the SINAD of the digital output exceeds a maximum limit. The difference between the desired RF signal level and the RF signal level of the adjacent channel interference signal is the adjacent channel interference rejection ratio.

Intermodulation Rejection

When an unwanted frequency's components are created in the receiver section as a result of the mixing of two or more RF interference signals, intermodulation distortion occurs. Sometimes the signals mix in the receiver and produce a signal that is the same frequency as the frequency that is supplied to the modulator.

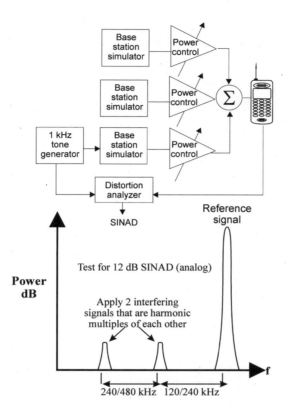

Figure 6.7, Intermodulation Rejection Measurements

Figure 6.7 shows how to test for inter-modulation rejection (analog). This test involves applying a reference signal to the receiver in the presence of two different radio signals. The frequencies of these other radio signals are selected so that their combination with the reference RF signal may produce frequency components that may interfere with the internal circuits of the mobile phone. The test involves increasing the RF signal level of the interference signals until the SINAD (analog) or FER (digital) levels exceed a maximum limit. This is the determining factor as to the amount of intermodulation rejection that can be tolerated. The difference between the desired RF signal level and the RF signal level of the interference signals is the intermodulation rejection ratio.

Spurious and Radiated Emissions

Similar to the mobile radio's transmitter circuits, a mobile radio's receiver assembly may produce a variety of radio signals within its chassis or housing. Although these RF signals should be contained within the mobile phone, some of these RF signals may radiate out of the case through connectors, unshielded cases and other parts of the mobile phone that do not successfully contain these unwanted transmissions.

There are two tests that are performed to ensure unwanted radio energy is not emanating from the mobile phone's receiver: conducted spurious emissions and radiated spurious emissions. Conducted spurious emissions are measured by connecting a spectrum analyzer to the antenna and scanning the frequency band to view unwanted RF signals. Radiated spurious emissions can emanate from different parts of the phone. For this test an antenna is connected to the spectrum analyzer and it is positioned at various points around the mobile phone to locate unwanted RF transmissions while the mobile phone is in the standby state.

Mobile Station Testing

The testing of mobile phones may be broken down into four categories: laboratory testing, environmental testing, production testing and field-testing.

Types of Testing

All facets of the CDMA phones and system operations are tested to explicit tolerances and protocols during the initial lab development. This warrants that the performance and protocol handling and its user features etc. are fully operational with a known RF environment.

Even before equipment designs are complete on working sections of the phone equipment, lab testing has already begun. This lab testing may discover that software changes are needed to correct problems that are found in the phone. It also guarantees that CDMA specifications and other special requirements, for a customer or distributor, are met. Some of the components tested include call origination, paging, short message service, system selection, channel reselection, intelligent roaming, soft hand-over, calling number identification, message waiting identification and other call processing features.

After all the lab testing is completed, the phone operation is tested to ensure it can meet the environmental requirements. The environmental tests include temperature, humidity levels, vibration (shaking, dropping), blowing dust and static discharge (shocking the phone) and may include added customers test requirements.

The next phase, after the lab and environmental testing is finished, is the production phase. The production phase also includes testing to authenticate and adjust the radio's performance to standards. Production testing is more limited compared to lab testing due to time issues. The less time for required testing, the large the number of units that can be tested by each piece of test equipment. In addition to the validation tests, production testing typically involves the automatic adjustment of electronic assemblies. This would include frequency and power adjustment.

Field testing occurs after the mobile phones have been introduced to the public and have failed operation. This type of testing is usually only used to ensure correct operation and not specific performance tolerances.

Test Interface Adapter

Test adapters enable the testing of mobile stations by connecting the mobile station directly to the test equipment. Standard test adapters make it practical to rely on accessing any sort of test modes. Without them, different test instructions would be required for the different manufacturers of different phones. Manufacturers have generally provided only Audio Input and Output and a RF jack or adapter so testing is done as much as possible directly in the customer's hand.

To reduce or eliminate the need for a test interface adapter, the loopback test mode is included in CDMA mobile phones. This allows the mobile phone to report back its signal conditions to a test set. Using the source and response measurement information, the test set can measure the performance of most parameters of the mobile phone without a test interface adapter.

Test Equipment

Field-testing is used to verify the correct operation of a mobile phone. Factory testing validates the tolerances and quality levels of most of the electronic circuits in the mobile phone. Production testing deals with the software and operation compatibility. Usually it is integrated into the manufacturer's computer system at the test facility, which allows external control. Test software is created to control automatic testing and adjustments. Each manufacturer has special test commands individualized for their own brand of radio. This allows internal access to certain segments that in turn permits the testing of individual parts separately

Precision test equipment is expensive and bulky. This has led to more portable test units that can be used in the field. Field test equipment typically is designed for specific technologies (such as CDMA or GSM) and sometimes has the capability to operate on 12 volts so they can be connected to the battery of test vehicles. Figure 6.9 shows a precision CDMA tester that can be moved to test sites in the field.

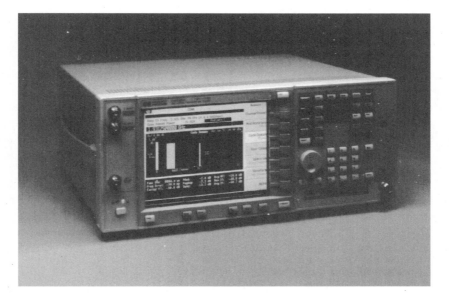

Figure 6.8 shows a production tester.
Source: Hewlett Packard

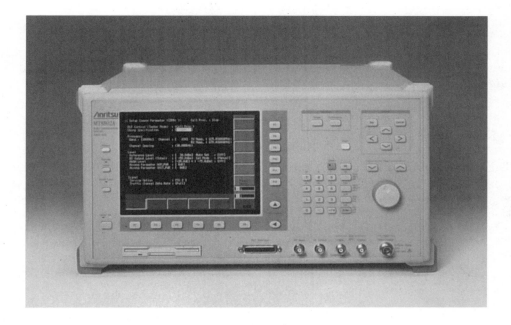

Figure 6.9, Field Tester
Source: Anritsu

Network Testing

Radio Propagation Testing

RF coverage quality and testing system operation are included in determining system verification. Signal quality is validated for individual cells first, and then adjacent cells are measured to determine system performance. System operation is verified by measuring the signal quality level at handoff, blockage performance, and the number of dropped calls.

The minimum percentage of area coverage is ensured through RF coverage area verification. Holes (dead spots) are located by measuring received signal strengths and interference levels. These holes are where the signal level goes below an acceptable level due to terrain and/or other obstructions. These problem areas can be corrected with an addition of another cell site or microcell or with the use of repeaters that augment the existing signal and focus the energy into the holes or dead spots.

System operation can be verified by registering the input levels obtained by a test mobile station when handoff and access occurs. Blocking probability can be calculated from the number of access attempts discarded by the system.

Network Equipment Testing

The operations administration and maintenance (OA&M) center contains the network equipment-testing portion. The CDMA network equipment interface communicates with the installed equipment in the network to ensure their correct operation.

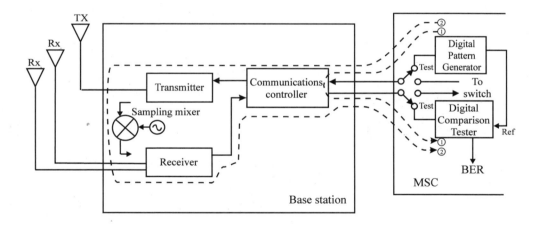

Figure 6.10, System Loopback Testing

CDMA networks are designed for reliability. There are many assemblies and communications links that are redundant so they can be placed out of service in the event of a failure. Networks have many automatic test systems to determine if a piece of equipment has failed. If the test system determines a failure or out of tolerance condition, the equipment may be disabled and tested.

Networks typically have test systems that can use parts of the network for loopback testing of assemblies throughout the network. Figure 6.10 shows a sample system loopback test system. In path 1, the network sends a test signal to the base and routes the test signal back to the system switch via the communications card. If the signal is successfully returned, the communications path and part of the communications card are verified as operational. For path 2, the test signal is directed to continue to the transmitter where it is sampled and sent back through the receiver. If the signal is successfully returned, the transmitter and receiver sections are also verified as operational. Various other network test configurations can be used to test and reconfigure network equipment.

Chapter 7

Marketing and Economics

Wireless telephone wholesale costs have dropped by approximately 20% per year over the past 7 to 10 years [1]. While the technology and mass production cost reductions for wireless telephones and systems are mature, new digital wireless telephones are more complex and usually do not have the large sales volume that promotes cost savings through mass production. System equipment costs for CDMA equipment must also compete against a mature, competitive analog cellular (e.g. AMPS, TACS, and NMT) equipment market, which already has the advantage of cost reductions due to large production runs.

The economic goal of a wireless network system is to effectively serve many customers at the lowest possible cost. The ability to serve customers is determined by the capacity of the wireless system. The key factors that primarily determine the capacity of the system is the size of the cell sites (ability to reuse frequencies or codes), the spectral efficiency of the radio channels (the number of users that can share each radio channel) and the number of radio channels that can be installed in each cell site. For any radio access technology, system capacity is increased by the addition of smaller cell site coverage areas, which allows more radio channels to be reused in a geographic area. If the number of cell sites remains constant, the efficiency of the radio access technology (e.g., the number of users that can share a single radio channel) and the ability to reuse frequencies (e.g. the number of radio channels in each cell site) determines the system capacity.

Wireless service providers usually strive to balance the system capacity with the needs of the customers. Running systems over their maximum capacity results in blocked calls to the customer, while running systems that have excess capacity results in the purchase of system equipment that is not required increases cost. Any wireless system (including analog cellular systems) can be designed for very high capacity use through the use of very small radio coverage areas. One of the key objectives of the new CDMA technology is to achieve cost-effective service capacity, using techniques such as digital voice compression, and voice activity.

Purchasing and maintaining wireless system equipment is only a small portion of the cost of a wireless system. Administration, leased facilities, and tariffs may play significant roles in the success of cellular systems.

The wireless marketplace is undergoing a change. New service providers that use digital land mobile radio technologies (e.g. iDEN and Tetra) are competing in the marketplace with new competitive features. This is likely to increase wireless services competition. Sales and distribution channels may become clogged with a variety of wireless product offerings. Advanced wireless digital technologies offer a variety of new features that may increase the total potential market and help service providers to compete. These new features may offer added revenue and provide a way to convert customers to a more efficient digital service. The same digital radio channels that provide voice services may offer advanced messaging and telemetry applications.

Marketing CDMA Phones

CDMA marketing programs have focused on converting existing analog subscribers to digital service by enticing new customers to purchase digital over analog for advanced services. The key marketing factors that may determine the success of CDMA includes the types of new services deployed by the network operator, system cost savings, pricing of voice and data service, mobile telephone cost, consumer confidence, new features, availability of equipment, and distribution channels.

Service Revenue Potential

At the end of 1998, there were over 286 million cellular telephone subscribers in the world [2] and over 23 million of these were CDMA cellular telephone users [3]. While the average cellular telephone bill is not much higher than the average wired residential telephone bill (approximately $50 per month), the amount of usage for a cellular telephone is approximately one tenth of residential usage.

The average cellular telephone bill in the US has declined eight to nine percent each year over the last five years [4]. This is not because the average charge per minute has decreased. It is because the amount of usage by new customers entering into the market is lower.

The rapid growth of the CDMA market is due to new system service areas in over 35 countries and the decreased price of CDMA telephones. In 1998, the total number of CDMA subscribers grew by 237% [5]. Because many new CDMA systems are starting, it is reasonable to assume a continued yearly growth of over 120% per year for most CDMA markets. In addition to the sale of CDMA handsets to consumers, it is possible to sell many times the number of handsets than those normally sold to people for non-human uses. For example, applications such as utility meter reading, vending machine management, vehicle tracking, environmental sensing and other applications have the potential for sales in excess of millions each worldwide. Figure 7.1 shows the world market for CDMA telephones.

The main revenue for wireless service providers is derived from providing telecommunications service. In 1998, a majority of the service revenue came from voice services. Digital wireless systems provide for increased service revenue that comes from a variety of sources such as advanced services and system cost reduction.

System Cost to the Service Provider

One of the advantages of digital service is to allow more customers to share the same system equipment. The cost to finance cellular system equipment costs (network equipment cost) account for approximately 10-15 % of the service provider's cost. Digital wireless systems can offer a reduction of approximately 60% of system equipment cost per customer. Some of the advanced features of digital wireless (such as authentication to reduce fraud) also provide for reductions in Operations, Administration, and Maintenance (OA&M) costs. The system cost reductions offered by digital wireless technology allow CDMA service providers to more effectively compete against other established wireless service providers.

Voice Service Cost to the Consumer

Over the past few years, the average cost of airtime usage to a cellular subscriber has not changed very much. To help attract subscribers to digital service, some cellular carriers have offered discounted airtime plans to high usage customers. This

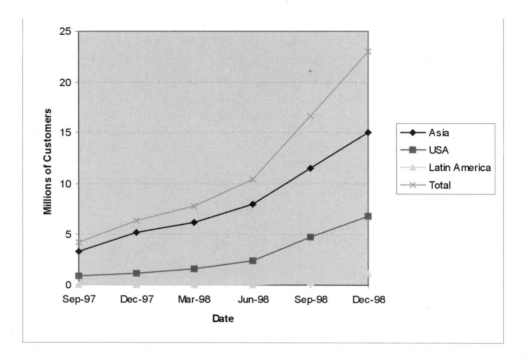

Figure 7.1, CDMA Customer Growth World Wide
Source: CDMA Development Group

discount provides a significant incentive to high usage customers from a competing system to convert to a system with CDMA technology.

Data Service Cost to the Consumer

There are two types of data services that are available to customers: continuous (called "circuit switched data") or brief packets (called "packet switched data"). Commonly, continuous data transmission airtime usage is charged at the same rate as voice transmission. Packet data transmission is often charged by the packet or by the total amount of data that has been transferred. When charging by the packet (approximately 30 bytes), the cost per packet ranges from approximately 1/5 cent to over fi cent per packet. When charging by the kilobyte of information, the cost ranges from approximately 7 cents to over $1 per kilobyte.

Wireless Telephone (Mobile Phone) Cost to the Consumer

In 1984-85, cellular mobile telephone prices varied from $2000-$2500 [6]. By 1991, a consumer could get a free cellular phone with the purchase of a hamburger at selected Big Boy Restaurants in the United States [7]. One of the primary reasons for the continued high growth of the cellular market is the declining terminal equipment costs and stable or reduced usage charges. In 1998, the wholesale price of digital wireless telephones was approximately 2 times higher than analog wireless telephones. Some CDMA service providers subsidize the sale of end-user telephones, which reduces the initial cost of the phone to the customer. Wireless service providers do not usually anticipate revenues from the sale of telecommunications equipment. Many of the service providers are not concerned with the profit for the sale of wireless telephone equipment because their goal is to gain monthly service revenue for usage of their system. Figure 7.2 shows the average retail price for handheld telephone cost in the United States over the past several years [8].

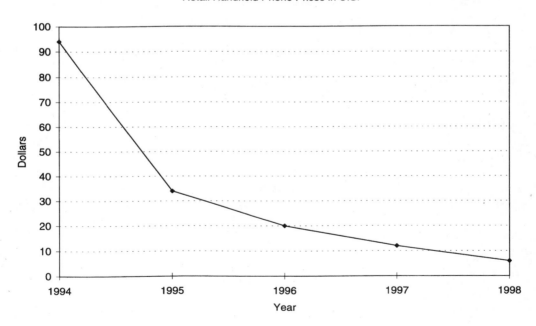

Figure 7.2, Retail Handheld Phone Prices in the United States
Source: Herschel Shosteck Associates, Wheaton, Maryland, USA

Consumer Confidence

To effectively deploy a new technology, the consumer must have confidence that the technology will endure. In some parts of the world, there are multiple digital technologies. These technologies offer different features, services, and radio coverage areas. Consumers must be willing to choose a technology that may not exist in future years. CDMA technology is deployed worldwide in over 35 countries. This allows customers to have confidence that the technology will remain available for many years.

New Features

Customers purchase mobile telephones and wireless service based on their own value system, which estimates the benefits they will receive. New features provide for new benefits to the consumer. These features can be used for product differentiation and to increase service revenue. Key new features available for CDMA systems include digital voice quality, voice privacy, messaging services, longer battery life and many others. These new features are used to persuade customers to convert from using an analog telephone or to pay extra for these new advanced services.

While the first analog cellular telephone weighed over 80 pounds and required almost all-available trunk space, the first digital wireless telephones were only slightly larger than their analog predecessors. The size of digital wireless telephones continues to be reduced as production volumes allow for custom integrated circuit development, which integrates the analog and digital processing sections. Digital wireless telephones should approach the same size and cost as analog cellular telephones over the next few years [9].

New features attract different types of customers. With advanced data capabilities, CDMA service providers are starting to focus its products, services, and applications on non-human applications such as telemetry or remote control. Without a change in focus, wireless service providers could become limited to the voice services market that may eventually reach saturation.

Churn

Churn is the percentage of customers that discontinue cellular service for any reason. Churn is usually expressed as a percentage of the existing customers that disconnect over a one-month period. Churn is often the result of natural migration (customers relocating) and the switching of service to other service providers. Because some wireless service providers contribute an activation commission incentive to help reduce the sale price of the phone, which can be a significant cost if the churn rate is high. The percentage of churn in North America over the last five years has remained relatively constant at approximately 2.8% per month [10].

Some CDMA carriers and their agents have gone to various lengths to reduce churn. This includes programming in lockout-code into the mobile equipment to avoid conversion to another carrier's service. The carrier can automatically enter programming lockout-codes (usually 4-8 digits) directly into some CDMA mobile telephones to keep the mobile telephone from reprogrammed with another carriers identification information. Wireless service providers sometimes require the customer to sign a service agreement, which as a rule requires them to maintain service for a minimum period of time (commonly one year). These service agreements have a penalty fee in the event the customer disconnects service before the end of the one-year period.

Availability of Equipment

The design and production of CDMA telephone equipment requires significant investment by a manufacturer. CDMA telephones are more complex, and portable digital wireless telephones are ordinarily larger and more expensive. Because CDMA systems are available in over 35 countries, there are many manufacturers. This has increased competition and reduced wholesale prices closer to the analog equivalent telephones (similar size and features).

The first cellular portable, introduced in 1984 by Motorola, weighed 30 oz and had approximately 30 minutes of talk time. In early August 1991, Motorola released the Micro-Tac Light which weighed 7.7 oz [11] and had 45 minutes of talk time. The development of advanced digital signal processing integrated circuits has allowed the production of CDMA telephones weighing less than 140 grams (5 oz).

Distribution and Retail Channels

Products produced by manufacturers are distributed to consumers via several distribution and retail channels. The types of distribution channels include: wholesalers, specialty stores, retail stores, power retailers, discount stores, and direct sales.

Wholesalers purchase large shipments from manufacturers and normally ship small quantities to retailers. Wholesalers will usually specialize in a particular product group, such as pagers and cellular phones.

Specialty retailers are stores that focus on a particular product category such as a cellular phone outlet. Specialty retailers know their products well and are able to educate the consumer on services and benefits. These retailers usually get an added premium via a higher sales price for this service.

Retail stores (retailers) provide a convenient place for the consumers to view products and make purchases. Retailers often sell a wide variety of products, but a salesperson may not have an expert's understanding of or be willing to dedicate the time to explain the features and cellular service options. In the early 1990s, mass retailers began selling cellular phones. Mass retailers sell a very wide variety of products at a low profit margin. The ability to sell at low cost is made possible by limiting the amount of sales time spent providing customer education on new features.

Power retailers specialize in a particular product group such as consumer electronics or office supplies. Power retailers look for particular product features that match their target market. Power retailers carry only a select group of products. Because there are only a few products for the consumer to select from, the demand for a single product is higher than if several different models were on display. This tends to increase sales for a particular product, which leads to larger quantity purchases and discounts for the power retailer.

Discount stores sell products at a lower cost than their competitors. They achieve this by providing a lower level of customer service. Because there is limited customer service, many of the wireless telephones sold in discount stores are pre-programmed or are debit (pre-paid) units.

Some wireless service providers employ a direct sales staff to service large customers. These direct sales experts can offer specialty service pricing programs. The sales staff may be well trained and regularly sell at the customer's location.

Distribution channels are commonly involved in the activation process. The application for cellular service usually takes 10 to 30 minutes, and programming of the mobile telephone for the consumer must be performed. In some cases, the handset programming can be accomplished using over the air programming. When the customer makes their first call, the call is automatically diverted to a CDMA service operator (called "Hotline"). The operator asks for the necessary information to activate the customer. The programming information (telephone number and feature set) is then sent to the customer via the CDMA radio channel. This information is automatically stored in the handset.

Because there are several new technologies and different models of phones, access to particular distribution channels is limited. In 1998, each retailer carried (stocked) approximately 3-4 different manufacturer's brands. Retailers can only dedicate a limited amount of shelf space for each product or service, which may limit the introduction of new digital products and services into the marketplace.

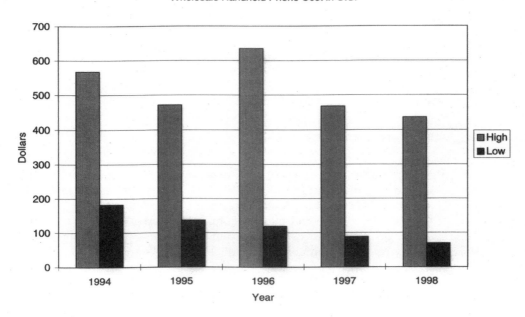

Figure 7.3, Wholesale Handheld Phone Cost in the United States
Source: Herschel Shosteck Associates, Wheaton, Maryland, USA

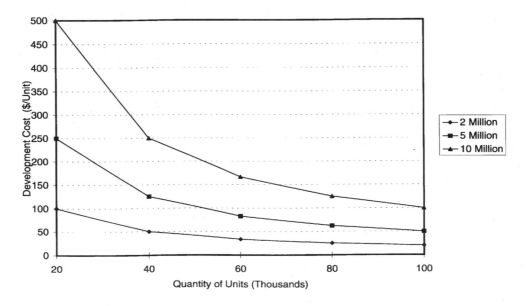

Figure 7.4, Mobile Telephone Development Cost

Wireless Telephone Costs

The initial digital cellular mobile telephones introduced in 1992 were approximately two times the cost of their analog equivalents. The average wholesale cost of analog mobile telephones (AMPS) has dropped from $307 in 1992 to approximately $104 in 1996 [12]. The higher cost of digital mobile telephones is due to the following primary factors: development cost, production cost, patent royalty cost, marketing, post-sales support, and manufacturer profit. Figure 7.3 shows the rapid decline of the wholesale cost of handheld wireless telephones in the United States.

Development Costs

Development costs are non-recurring costs that are required to research, design, test, and produce a new product. Unlike well-established FM technology, non-recurring engineering (NRE) development costs for CDMA telephones can be high, due to the added complexity of digital design. Several companies have spent millions of dollars developing digital wireless products. Figure 7.x shows how the non-recurring development cost per unit varies as the quantity of production varies from 20,000 to 100,000 units. Even small development costs become a significant challenge if the volume of production of the digital wireless telephones is low (below 20,000 units). At this small production volume, NRE costs will be a high percentage of the wholesale price.

The introduction of a new technology presents many risks in terms of development costs. Some development costs that need to be considered include: market research; technical trials and evaluations; industrial, electrical, and software design; prototyping; product and government type approval testing; creation of packaging, brochures, user and service manuals; marketing promotion; sales and customer service training; industry standards participation; unique test equipment development; plastics tooling; special production equipment fabrication; and overall project coordination.

When a new product is created which is not a revision of an existing product using readily available components sometimes compromises a cost-effective design. Cost effective design is achieved by integrating multiple assemblies into a custom chip or hybrid assembly. Custom integrated circuit chip development is used to integrate many components into one low-cost part. Excluding the technology development effort, the custom Application Specific Integrated Circuit (ASIC) development usually requires a development setup cost that ranges from $250,000 to $500,000. There may be more than one ASIC used in a digital wireless telephone.

Cost of Production

The cost to manufacture a mobile telephone includes the component parts (bill of materials), automated factory assembly equipment, and human labor. Digital wireless telephones are more complex than the analog units. A digital mobile telephone is composed of a radio transceiver and a digital signal processing section. The primary hardware assemblies that affect the component cost for digital mobile telephones are Digital Signal Processors (DSPs) and radio frequency assemblies. A single DSP, and several may be used, may cost between $7 and $28 [13]. The Radio

Frequency (RF) assemblies used in CDMA telephones require fast switching frequency synthesizers as compared to analog phones. In 1998, these RF components cost approximately $10-15. Other components that are included in the production of a mobile telephone include printed circuit boards, integrated circuits and electronic components, radio frequency filters, connectors, plastic case, a display assembly, a keypad, a speaker and microphone, and an antenna assembly. In 1998, the bill of materials (parts) for a digital mobile telephone was approximately $90 [14].

The assembly of wireless telephones requires a factory with automated assembly equipment. Each production line can cost between two and five million dollars. Regularly, one production line can produce a maximum of 500-2,000 units per day (150,000-600,000 units per year). The number of units that can be produced per day depend on the speed of the automated component insertion machines and the number of components to be inserted. Normally, production lines are often shut down one day per week for routine maintenance and two weeks per year for major main-

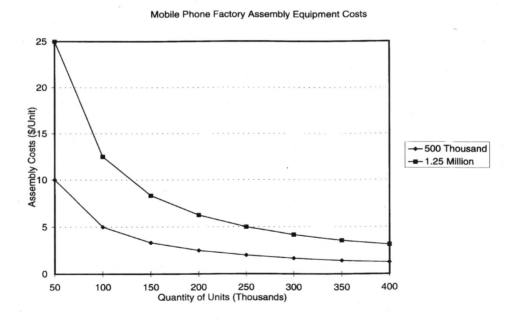

Figure 7.5, Mobile Phone Factory Assembly Equipment Cost

tenance overhauls which leaves about 300 days per year for the manufacturing line to produce products. Between interest cost (10-15% per year) and depreciation (10-15% per year), the cost to own such equipment is approximately 25% per year. This results in a production facility overhead of $500 thousand to $1.25 million per year for each production line. Figure 7.5 shows how the cost per unit drops dramatically from approximately $10-25 per unit to $1-3 per unit as volume increases from 50,000 units per year to 400,000 units per year.

While automated assembly is used in factories for the production of mobile telephones, there are some processes that require human assembly. Efficient assembly of a mobile telephone in a modern factory requires 1/2 to 1 hour of human labor. The amount of human labor is a combination of all workers involved with the plant, including administrative workers and plant managers. The average loaded cost of labor (wages, vacation, insurance) varies from approximately $20 to $40 per hour, which is based on the location of the factory and the average skill set of human labor. This results in a labor cost per unit that varies from $10-40. Because digital wireless telephones may have more parts to assemble due to the added complexity, the labor cost may increase.

Patent Royalty Cost

Another significant cost factor to be considered is patent royalties. Cellular technology was originally developed and patented by AT&T [15]. To the author's knowledge, AT&T has never requested a single royalty payment for this fundamental technology, which is not the case for the new digital standards. Several companies have disclosed that they believe they have some proprietary technology that is required to implement some features defined in the CDMA standard specifications [16]. Some large manufacturing companies exchange the right to use their patented technology with other companies that have patented technology they want to use. Patents from other companies that may be desirable or essential to implement the standard specifications may not have been discovered or disclosed.

Marketing Cost

The marketing cost, which is included in the wholesale cost of the wireless telephone, includes a direct sales staff, manufacturer's representatives, advertising, trade shows, and industry seminars.

Wireless telephone manufacturers ordinarily dedicate a highly paid representative or agent for key customers. Much like the sales of other consumer electronics product, manufacturers employ several technical sales people to answer a variety of technical questions prior to the sale.

There are wireless telephone manufacturers that use independent distributors to sell their products. This practice is more prevalent for smaller, lesser-known manufacturers who cannot afford to maintain dedicated direct sales staff. These representatives commonly receive up to four percent of the sales volume for their services.

Advertising programs used by the wireless telephone manufacturers involve broad promotion for brand-recognition and advertisements targeted for specific products. The typical advertising budget for mobile telephone manufacturers varies from approximately three to six percent. The budget for brand recognition advertising normally ranges from less than one percent to over four percent. Product-specific advertising is often performed through co-operative advertising. Co-operative advertising is paid from the manufacturer to a distributor or retailer. The amount paid commonly varies from 2% to 4% of the cost of the products sold to the distributor. For the distributor or retailer to receive the co-operative commission, they must meet the manufacturers advertising requirements. This approach allows distributors and retailers to determine the best type of advertising for their specific markets.

Cellular system equipment manufacturers ordinarily exhibit at trade shows 3 to 4 times per year. Trade show costs are high. Cellular mobile telephone manufacturers exhibiting at trade shows normally have large trade show booths, gifts, and theme entertainment. Medium to large hospitality parties at the trade shows is also common. Wireless telephone manufacturers often bring 15 to 40 sales and engineering experts to the trade shows to answer distributor questions.

To help promote the industry and gain publicity, wireless telephone manufacturers participate in a variety of industry seminars and associations. The manufacturers regularly have a few select employees who write for magazines and speak at industry seminars. All of these costs and others result in a combined estimated marketing cost for mobile telephone manufacturers of 10 to 15 percent of the wholesale selling price.

Post Sales Support

The sale of cellular mobile telephones involves a variety of costs and services after the sale of the product (post sales support), including warranty servicing, customer service, and training. A customer service department is required for handling distributor and customer questions. Because the average customer for a wireless telephone is not technically trained in radio technology, the amount of non-technical questions can be significant. Distributors and retailers require training for product feature operation and servicing. The post sales support cost for wireless telephones is usually between 4 to 6 percent.

Manufacturer's Profit

Manufacturers must make a profit as an incentive for manufacturing products. The amount of profits a manufacturer can make as a rule depends on the risk involved with the manufacturing of products. As a general rule, the higher the risk, the higher the manufacturers' profit margin.

The wireless telephone market in the early 1990's became very competitive due to the manufacturers' ability to reduce cost through mass production. To effectively compete, manufacturers had to invest in factories and technology, which increased the risk and the required profit margin. In 1995, the estimated gross profit in the wireless telephone manufacturing industry was 15 to 35 percent [17].

System Equipment Costs

The cost for wireless system equipment includes the following primary factors: development cost, production cost, patent royalty cost, marketing, post sales support, and manufacturer profit.

Development Costs

The wireless network system equipment development costs are much higher than wireless telephone development costs. When a completely new technology is introduced, wireless network system development costs can exceed $500 million because the complexity of an entire wireless system is significantly greater than the complexity of a mobile telephone. Thus more testing and validation is required. This high investment would limit most manufacturers from producing products for the new technology.

While the base station radios perform similar to a wireless telephone, the coordination of all the wireless telephones involves many electronic subsystems. Additional assemblies include communication controllers in the base station and switching center, scanning locating receivers, communication adapters, switching assemblies, and large databases to hold subscriber features and billing information. All of these assemblies require hardware and very complex software.

Unlike mobile telephones, when a cellular system develops a problem, the entire system can be affected. New hardware and features require extensive testing. Testing cellular systems can require thousands of hours of labor by highly skilled professionals. Introducing a new technology is much more complex than simply adding a new feature.

Cost of Production

The cost to manufacture a wireless network system includes the component parts, automated factory equipment, and human labor. Because the number of subscribers that share a single CDMA radio channel can exceed 500 (not all users access the system at the same time), the quantity of wireless network system assemblies produced is much smaller than the number of wireless telephones. Setting up automated factory equipment is time consuming. For small production runs, much more human labor is used in the production of assemblies because setting up the automated assembly is not practical. The production of system equipment involves a factory with automated assembly equipment for specific assemblies. However, because the number of units produced for system equipment is normally much smaller than wireless telephones, production lines used for cellular system equipment are often shared for the production of different assemblies, or remains idle for periods of time.

With over 35 countries using CDMA systems, the demand for CDMA system equipment is increasing exponentially. This increased demand allows for larger production runs, which reduce the average cost per unit. Large production runs also permit investment in cost-effective designs, such as using Application Specific Integrated Circuits (ASICs) to replace several individual components.

The maturity of digital technology is promoting cost reductions through the use of cost-effective equipment design and low-cost commercially available electronic components. In the early 1990s, many technical system equipment changes were required due to changes in radio specifications. Manufacturers had to modify their equipment based on field test results. For example, complex echo cancelers were required due to the long delay time associated with digital speech compression. Manufacturers ordinarily did not invest in cost-effective custom designs because of the rapid changes. As the technology has matured, the investment in custom designs is possible with less risk. In the early 1990s, it was also unclear which digital technologies would become commercially viable, which limited the availability of standard components. Today, the success of digital systems has created a market of low-cost digital signal processors and RF components for digital cellular systems.

Like the assembly of wireless telephones, the assembly of system radio and switching equipment involves a factory with automated assembly equipment. The primary difference is the smaller production runs, multiple assemblies, and more complex assembly.

The number of equipment units that are produced is much smaller than the number of mobile telephones produced because each CDMA radio channel produced can serve over 500 subscribers. The result is much smaller production runs for wireless network system equipment. While a single production line can produce a maximum of 500-2,000 assemblies per day [18], several different assemblies for radio base stations are required. A change in the production line from one assembly process to another can take several hours or several days. Wireless system radio equipment requires a variety of different connectors, bulky RF radio parts, and large equipment case assemblies. Due to the low-production volumes and many unique parts, it is not usually cost effective to use automatic assembly equipment. For unique parts, there are no standard automatic assembly units available. Because of this more complex assembly and the inability to automate many assembly steps, the amount of human labor is much higher than for wireless telephones.

Each automated production line can cost two to five million dollars. The number of units that can be produced per day on a single production line varies depending on the speed of the automated component insertion machines, the number of components to be inserted, the number of different electronic assemblies per equipment. In addition, it takes additional time to change/setup the production line for different assemblies. If it is to be assumed there are four electronic assemblies per base station radio equipment (e.g., controller, RF section, baseband/diagnostic processing section, and power supply), the automated production cost for base station equipment should be over four times that of mobile telephones.

Figure 7.6 shows how the production cost per unit drops dramatically from approximately $400-1,000 per unit to $50-125 per unit as the volume of production increases from 5,000 units per year to 40,000 units per year. This chart assumes produc-

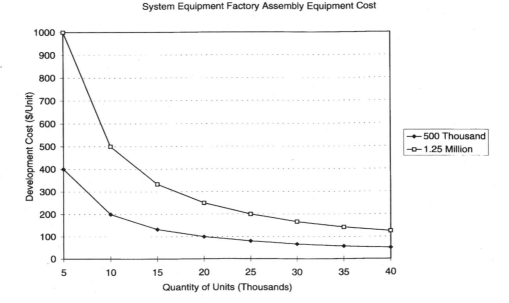

Figure 7.6, System Equipment Factory Assembly Equipment Cost

tion cost is four times that of wireless telephones due to the added complexity and the use of multiple assemblies.

While automated assembly is used in factories for the production of wireless telephones, there are some processes that require human assembly. Efficient assembly of base station units in a modern factory requires between 5 and 10 hours of human labor. The amount of human labor includes all types of workers from administrative workers to plant managers. The average loaded cost of labor (wages, vacation, insurance) varies from approximately $20-40 per hour, which is based on the location of the factory and average workers skill set. The resultant labor cost per unit varies from $100-400.

Patent Royalty Cost

There are fewer manufacturers that produce wireless network system equipment due to the fact that the use of many different technologies is involved. Large manufacturers have a portfolio of patents that are commonly traded. Cross licensing is common and tends to reduce the cost of patent rights. When patent licensing is required, the patent costs are sometimes based on the wholesale price of the assemblies in which the licensed technology is used.

Marketing Cost

The marketing costs that are included in the wholesale cost of wireless system equipment include a direct sales staff, sales engineers, advertising, trade shows, and industry seminars. Wireless system manufacturers often dedicate several highly paid representatives for key customers. Wireless system sales are much more technical than the sale of wireless telephones. Manufacturers employ several people to answer a variety of technical questions prior to the sale.

Advertising used by the cellular system equipment manufacturers involves broad promotion for brand recognition and advertisements targeted for specific products. The budget for brand recognition advertising is commonly small and is targeted to specific communication channels because the sale of wireless system equipment involves only a small group of people who usually work for a wireless service provider. Product-specific advertising is also limited to industry specific trade journals. Much of the advertising promotion of wireless system equipment occurs at trade shows, industry associations, and client sales presentations. The advertising budget for wireless system equipment manufacturers is regularly less than two percent.

Wireless system manufacturers also exhibit at trade shows normally 3 to 4 times per year. The trade show costs are sometimes much higher than the trade show costs for wireless telephone manufacturers. Wireless system equipment manufacturers exhibiting at trade shows often have large hospitality parties that sometimes entertain thousands of people. Wireless system manufacturers often bring 60-100 sales and engineering experts to the trade shows to answer customer questions.

To help promote the industry and gain publicity, wireless system manufacturers participate in many industry seminars and associations. These manufacturers use trained experts to present at industry seminars. All of these costs and others result in an estimated marketing cost for system equipment manufacturers of approximately 8-10% of the wholesale selling price.

Post-Sales Support

The sale of wireless systems involves a variety of costs and services after the sale of the product. This includes warranty servicing, customer service, and training. A 7-day x 24-hour customer service department is required for handling customer questions. Customers require a significant amount of training for product operation and maintenance after a system is sold and installed. The post-sales support costs for wireless system equipment is ordinarily 3-5%.

Manufacturer Profit

Standardization of systems and components, particularly CDMA, has led to a rapid drop in the wholesale price of system equipment. While the increased product volume of wireless system equipment has resulted in decreased manufacturing costs, the gross profit margin for wireless system equipment has decreased. The estimated gross profit in the wireless telephone manufacturing industry is 10-15% [19].

Network Capital Costs

The wireless service provider's investment in network equipment includes cell sites, base station radio equipment, switching centers, and network databases. One of the primary objectives of the new technologies was to decrease the network cost per customer (capital cost per customer) which was made possible because the new technologies can serve more customers with less physical equipment.

In theory, existing analog cellular technology can serve an almost unlimited number of subscribers in a designated area by replacing large cell site areas with many Microcells (small cell coverage areas). However, expanding the current analog systems in this way increases the average capital cost per subscriber due to the added cost of increasing the number of small cells and interconnection lines to replace a single large cell. For example, when a cell site with a 15-km radius is replaced by cell sites with a 1/2 km radius, it will take over 700 small cells to cover the same area.

One of the reasons that digital cellular technologies were developed was to allow for cost-effective capacity expansion. Cost-effective capacity expansion results when existing cell sites can offer more communication channels, which allows more customers to be served by the same cell site. As systems based on such new technologies expand, the average cost per subscriber decreases.

Cell Site

The cell site is composed of a radio tower, antennas, a building, radio channels, system controllers, and a backup power supply. The cell site radio tower is commonly 100-300 feet tall. The cost ranges between 30 thousand and 300 thousand dollars. While some of the largest towers can cost $300 thousand, an average cost of $70 thousand is typical because, as systems expand, smaller towers can be used.

Many cell sites can be located on a very small area of land. Land is either purchased or leased. In some cases, existing tower space can be leased for $500-$1,000 per month. If the land is purchased, the estimated cost of the land is approximately $100 thousand.

A building on the cell site property is required to store the cell site radio equipment. This building must be bullet proof, have climate control, and various other non-standard options. The estimated building cost is $40 thousand.

Item	Cost (000's)
Radio Tower	$70
Building	$40
Land	$100
Install Comm Line	$5
Construction	$50
Antennas	$10
Backup Power Supply	$10
Total	$285

Table 7.1, Estimated Cell Site Capital Cost Without Radio Equipment

Cell sites are not usually located where high-speed telephone communication lines are available. Typically, it is necessary to install a T1 or E1 trunk communications line to the cell-site that is leased from a local telephone company. If a microwave link is used in place of a leased communication line, the communications line installation cost will be applied to the installation of the microwave antenna. The estimated cost of installing a T1 or E1 communications line is approximately $5 thousand dollars.

The land where the cell site is to be located must be cleared, foundations poured, fencing installed, building and tower installed. A construction cost of $50 thousand is estimated. Table 7.1 shows the estimated cost for a typical cell site without the radio equipment.

In addition to the tower and building cost, radio equipment must be purchased. Each CDMA cell site usually has several radio transceivers. To determine the total number of subscribers that can be serviced by a cell site, the number of radio transceivers is multiplied by the number of subscribers that can be serviced by each radio transceiver.

After the total investment of each cell site is determined, the cell site capital cost per customer can be determined by dividing the total cell site cost by the number of subscribers that will share the resources (cell sites).

Table 7.2 shows a simplified sample of system equipment costs as digital technology evolves. In column 1, we see analog technology that supports one voice channel per carrier. If the average cost of a radio channel for analog technology is $10,000 (this includes the line adapters and a proportion of the controller cost) and 51 radio channels are installed per cell site, the total radio equipment cost is approximately $510,000. The tower and building cost is added to this bring the total cell site cost to approximately $795,000. Three of the radio channels are used for control channels, which leaves 48 channels available for voice communications. Because each subscriber will only access the cellular system for a few minutes each day, approximately 25 subscribers (customers) can share the service of a single radio channel. This brings the total number of subscribers that can be added for each analog cell site to 1200. If the average cost of an analog cell site is $795,000, this brings the average cell site cost per analog customer to $633.

Column 2 in table 7.2 shows the average cost for a typical full rate CDMA system. If the average cost of a radio transceiver for CDMA technology is $20,000 (this includes the line adapters and a higher proportion of the controller cost than analog) and 15 radio transceivers are installed per CDMA cell site, the total radio equipment cost is approximately $300,000. The tower and building cost is added to

	Analog	CDMA full rate
Cost per RF Radio Channel	$10,000	$20,000
Number of Radio Channels per Cell Site (3 sector)	51 (3 for control)	15
Total Radio Channel Cost	$510,000	$300,000
Tower and Building Cost	$285,000	$285,000
Total Cell Site Cost	$785,000	$585,000
Number of Voice Paths per Radio Channel	1	20
Number of Voice Paths per Cell Site	48	300
Number of Subscribers per Voice Channel (loading)	25:1	25:1
Number of Subscribers per Cell Site	1,200	7,500
Cell Site Capital Cost per Customer	$663	$78

Table 7.2, Cell Site Capital Cost per Subscriber

this to bring the total cell site cost to approximately $585,000. Each CDMA radio carrier provides approximately 20 communication channels (64 channels maximum although some are used for control and soft handoff). This brings the total number of available communication channels (for voice) to 300. Because each subscriber will only access the cellular system for a few minutes each day, approximately 25 subscribers (customers) can be added for each communications (voice) path. This brings the total number of subscribers that can be added for each CDMA cell site to 7500. If the average cost of a CDMA cell site is $585,000, this brings the average cell site cost per CDMA customer to $78.

The ability of radio channels to serve more customers through one RF piece of equipment reduces the number of required RF equipment assemblies, power consumption, and system cooling requirements. Sharing a channel in this way (multiplexing) reduces cell site size and backup power supply (generator and battery) requirements, and ultimately, cost.

Repeaters

Repeaters can be used to extend the radio coverage area without incurring many of the costs associated with cell sites. Repeaters receive a radio signal from a donor cell, amplify the signal and retransmit the signal to a new geographic territory.

Although repeaters do not increase the capacity of a cellular system, they can be used to extend the radio coverage area at a significantly lower cost than adding cell sites. Table 7.3 below shows that the total cost of cell sites includes initial capital investment and recurring costs. Repeaters can reduce the initial equipment costs through the elimination of communications controllers and network switching requirements. In addition, communication lines do not need to be installed for repeaters.

The operational cost reductions for repeaters are dramatically lower than for cell sites. The monthly leased line cost is eliminated, utilities are reduced because of the lower environmental cooling requirements. Maintenance and software licensing costs are usually lower.

Initial Cost	Base Station	Microcell	Repeater
Equipment	$150,000	$50,000	$30,000
Switching	$60,000	$20,000	$0
Site Acquisition	$40,000	$40,000	$40,000
Tower	$120,000	$120,000	$120,000
RF Engineering	$25,000	$25,000	$25,000
T1/E1 Install	$5,000	$5,000	$0
Total Initial Cost	**$400,000**	**$260,000**	**$215,000**
Annual Operational (Recurring Cost)			
T1/E1 Line Lease	$24,000	$24,000	$0
Utilities	$1,800	$600	$600
Maintenance	$2,100	$700	$700
Software	$2,100	$700	$700
Total Operating Cost	**$30,000**	**$26,000**	**$2,000**

Table 7.3, Total Cost of Cell Site Ownership (Case Study)
Source: Repeater Technologies

Mobile Switching Center

Cell sites must be connected to an intelligent switching system (called the "switch"). An estimate of $25 per subscriber is used for the cellular switch equipment and its accessories, based on one Mobile Switching Center (MSC) costing $2.5 million that can serve up to 100,000 customers.

The switching center must be located in a long-term location (10-20 years), usually near a Public Switched Telephone Network (PSTN) central office switch connection. The building contains the switching and communication equipment. Commonly, customer databases (HLR and VLR) are located in the switching center facility. The switching center software and associated cellular system equipment commonly contain basic software that allows normal mobile telephone operation (place and receive calls). Special software upgrades that allow advanced services are available at additional cost.

Operational Costs

The costs of operating a cellular system include leasing and maintaining communication lines, local and long distance tariffs, billing, administration (staffing), software licensing, maintenance, and cellular fraud. The operational cost benefits of installing digital equipment includes a reduction in the total number of leased communication lines, a reduction in the number of cell sites, a reduction in maintenance costs, and a reduction of fraud due to advanced authentication procedures.

Leasing and Maintaining Communications Lines

Leased communication lines between radio towers, or must connect cell sites by installing and maintaining microwave links between them. The typical cost for leasing a 24-channel line between cell sites in the US in 1998 ranged from $300 to $550 per month ($500 typical) [20]. The average cost of leasing a 30 channel E1 line in Europe ranged from $450 to over $2000 per month ($800 typical) [21]. Leasing cost for communication lines depends on the distance of connection points and guaranteed grade of service. The greater the distance between the connection points, the higher the leased line cost. Because of increased competition in the telecommunications area by cable companies and competitive local exchange carriers offering high speed digital leased lines, it is anticipated that the average cost of leased E1 and T1

Service	T1 Line Cost per Month	Number of Channels per Line	Usage per Month (min)	Customers per Channel	Customers per Line	Total Cost per Month
Cellular	$500	24	100	25	600	$0.83
LEC (residential)	$500	24	1000	5	120	$4.17
Office	$500	24	2000	2.5	60	$8,33

Table 7.4, T1 Monthly Communications Line Cost

Service	T1 Line Cost per Month	Number of Channels per Line	Usage per Month (min)	Customers per Channel	Customers per Line	Total Cost per Month
Cellular	$800	30	100	25	750	$1.07
LEC (residential)	$800	30	1000	5	150	$5.33
Office	$800	30	2000	2.5	75	$10.66

Table 7.5, E1 Monthly Communications Line Cost

lines will continue to decrease. Installing microwave radio equipment can eliminate the monthly cost of leased lines. The cost of microwave radio links between 2 cell sites ranges from approximately $20 thousand to $100 thousand.

Similar to the sharing of the resources of a radio carrier, the number of subscribers that can share the cost of a communication line (loading of the line) varies with the type of service. For cellular-like subscribers who ordinarily use the phone for two minutes per day, approximately 600 customers can share a T1 (25 subscriber per voice path x 24 voice paths per communication line) or 750 customers per E1 (25 subscribers per voice path x 30 voice paths per communication line). This can be compared to residential-type service where customers use the phone for approximately 30 minutes per day. For residential service, approximately 120 customers can be loaded onto a T1 or 150 per E1. The average office customers use the phone for approximately 60 minutes per day. For office usage, approximately 60 customers can be loaded onto a T1 or 75 for E1.

The monthly cost per subscriber is determined by dividing the monthly cost by the total number of subscribers. Tables 7.4 and 7.5 show the estimated monthly cost for interconnection charges. The estimated monthly cost is based on 100% use of the communication lines. If the communication lines are not used fully (it is rare that communication lines are used at full capacity), the average cost per line increases.

Digital signal processing for all the proposed technologies allow for a reduction in the number of required communications links through the use of sub-rate multiplexing. Sub-rate multiplexing allows several users to share each 64 thousand bit per second (kb/s) communications (DS0/PCM) channel. This is possible because digital cellular voice information is compressed into a form much smaller than the existing communication channels. If 13 kb/s speech information is sub-rate multiplexed, up to 4 voice channels can be shared on a single 64 kb/s channel, which can reduce the cost of leased lines significantly.

Local and Long Distance Tariffs

Telephone calls in cellular systems are often connected to other local and long distance telephone networks. When cellular systems are routed to the existing public telephone customers, they are normally connected through the local wired telephone network. The local telephone company regularly charges a small monthly fee and several cents per minute (approximately 3 cents per minute) for each line connected to the cellular carrier. Because each cellular subscriber uses their mobile telephone for only a few minutes per day, the cellular service provider can use a single connection (telephone line) to the PSTN to service hundreds of subscribers.

In the United States and other countries that have separate long-distance service providers, when long distance service is provided through a local telephone company (LEC), a tariff is paid from its cellular service provider to the local exchange company (LEC). These tariffs can be up to 45% of the charges for each minute of long distance service. Due to government regulations limiting the bundling of local and long-distance service, it is necessary for some cellular service providers to separate their local and long-distance service. Recent regulations may permit cellular carriers to bypass the LEC and save these tariffs.

Billing Services

Cellular systems exist to provide services and collect revenue for those services. Billing involves gathering and distributing billing information, organizing the information, and invoicing the customer.

As customers initiate calls or use services, call detail records are created. These records may be provided in the customer's home system or a visited system. Each billing record contains details of each billable call, including who initiated the call, where the call was initiated, the time and length of the call, and how the call was terminated. Each call record contains approximately 100-200 bytes of information [22]. If the calls and services are provided in the home system, the billing records can be stored in the company's own database. If they are provided in a visited system, the billing information must be transferred back to the home system. In the mid-1980s, cellular systems were not interconnected, which required the use of a magnetic tape to transfer records standard Automatic Message Accounting (AMA) format. Today, billing records are regularly sent directly to a clearinghouse company to accumulate and balance charges between different cellular service providers.

With the introduction of advanced services, billing issues continue to become more complicated. The service cost may vary between different systems. To overcome this difficulty, some service providers have agreed to bill customers at the billing rate established in their home system.

Each month, billing records must be totaled and printed for customer invoicing, invoices are mailed, and checks that are received are posted. The cost for billing services ranges approximately from $1 to $3 per month. Billing cost includes routing and summarizing billing information, printing the bill, and the cost of mailing. To help offset the cost of billing, some wireless service providers have started to bundle advertising literature from other companies along with the invoice. To expedite the collection, wireless service providers may offer direct billing to bank accounts or charge cards.

Activation Commissions (Equipment Subsidies)

Some service providers provide an activation subsidy (commission) to a retailer when a consumer purchases a mobile phone. This activation subsidy reduces the barrier of the high purchase price of a mobile handset from a consumer's decision to purchase a mobile phone. During 1997, activation commissions paid to retailers in the United States ranged from approximately $50 - $300 per handset. Figure 7.7 demonstrates the average commissions to the retailers from the wireless carriers.

To help ensure the recovery of the activation commission (which can be several hundred dollars US), the activation subsidy is ordinarily paid only if the customer signs a long term service agreement (normally 1-year minimum). The wireless telephone service activation subsidy and the type of distribution channels used usually affects the retail price paid by the consumer.

Software Licensing

The software that is used by the switching system and base stations to operate a cellular system and provide for advanced features is not usually owned by the carrier. The software is licensed from the manufacturer. The amount of cost of licensing primarily depends on the number of access lines and base stations. Software licensing in the US for base stations is approximately $2,000 per year [23].

Operations, Administration, and Maintenance (OA&M)

Running a wireless service company requires people with many different skill sets. Staffing requirements include executives, managers, engineers, sales, customer service, technicians, marketing, legal, finance, administrative, and other personnel to

Digital Activation Commissions to Retailers from Wireless Carriers

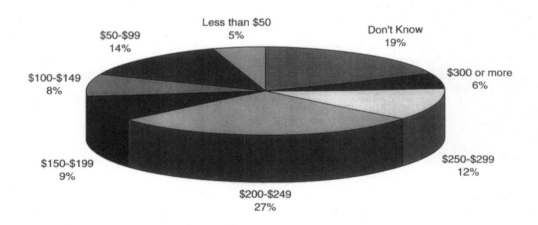

Figure 7.7, Digital Activation Commissions to Retailers

support vital business functions. The present staffing for local telephone companies is approximately 35 employees for each 10,000 customers. Wireless telephone companies have approximately 20-25 employees per 10,000 customers, and paging companies employ approximately 10 employees per 10,000 customers [24]. If we assume a loaded cost (salary, expenses, benefits, and facility costs) of $40,000 per employee, this results in a cost of $3.33 to $11.66 per month per customer ($40,000 x (10-35 employees) /10,000 customers/12 months).

Maintaining a wireless system requires calibration, repair, and testing. System growth involves frequency planning, testing, and repair. When a frequency plan is changed, manual tuning of 15-20 radio channels per cell site is required. Large urban systems may have over 400 cell sites. Because it is desirable to perform frequency re-tuning in the same evening, some wireless service providers borrow technicians from neighboring systems when their small staff cannot perform this task. The CDMA system has advanced features that simplify or automate radio frequency planning. This reduces OA&M cost.

During the year, the geographic characteristics of a wireless system change (e.g., leaves fall off trees). This changes the radio coverage areas. Usually, a wireless carrier tests the radio signal strength in its entire system at least four times per year. Testing involves having a team of technicians drive throughout the system and record the signal strength.

Maintenance and repair of wireless systems is critical to the revenue of a wireless system. In large systems, staffs of qualified technicians are hired to perform routine testing. Smaller wireless systems often have an agreement with another wireless service provider or a system manufacturer to provide these technicians when needed. Wireless systems have automatic diagnostic capabilities to detect when a piece of equipment fails. Most wireless systems have an automatic backup system, which can provide service until the defective assembly is replaced.

Land and Site Leasing

In rural areas, exact locations for cell site towers are not required. The result is that land leasing is not a significant problem. In urban areas and as systems mature, more exact locations for cell sites are required. This results in increased land-leasing costs. By using a more efficient RF technology, one cell site can be used to serve more channels, which limits the total number of required cell sites.

Land leasing is as a rule a long-term lease for very small portions of land (40-200 square meters) for approximately 20 years or longer. The cost of leasing land is dependent on location. Premium site locations such as sites on key buildings or in tunnels can exceed the gross revenue potential of the cell site.

Another leasing option involves leasing space on an existing radio tower. Site leasing on an existing tower is approximately $500/month. Site leasing eliminates the requirement of a building and maintaining a radio tower.

Cellular Fraud

It is estimated that cellular fraud in the United States during 1994 was in excess of $460 million [25]. This was approximately three percent of the $14 billion yearly gross revenue received [26]. Each of the new cellular standards has an advanced authentication capability, which limits the ability to gain fraudulent access to the cellular network.

The type of cellular fraud seen has changed over the years. Initially, cellular fraud was subscription fraud. With advances in technology available to distributors of modified equipment, cellular fraud has changed to various types of access fraud. Access fraud is the unauthorized use of cellular service by changing or manipulating the electronic identification information stored inside of a mobile telephone.

Subscription fraud occurs when a person using false identification registers a cellular phone. After the required documentation is provided, the cellular service provider often provides unlimited service. When the bill is unpaid, the fraudulent activity is determined and service is disconnected. Some cellular service providers now require valid identification and credit checks prior to service activation, which reduces subscription fraud.

In the mid-1980s, roamer fraud was possible. Roamer fraud occurs when a mobile telephone is programmed with an unauthorized telephone number and home system identifier so that it looks like a visiting customer. Because some of the cellular systems in the mid-1980s were not directly connected to each other, these systems could not immediately validate the visiting customer. CDMA systems are interconnected to provide validation (authentication) of the subscriber which limits (or eliminates) roamer fraud.

Most cellular fraud attempts can be detected and blocked by the use of mobile telephone authentication information. Authentication is a process of using previously stored information to process keys that are transferred via the radio channel. Because the secret information is processed to create a key, the security information is not transferred on the radio channel. The secret information stored in the wireless telephone can be changed at random either by manual entry, by the customer, or by a command received from the wireless system. Authentication is supported in all of the digital technology specifications.

References:

[1]. Herschel Shosteck, "The Retail Market of Cellular Telephones", Herschel Shosteck Associates, Wheaton, Maryland, USA, 1st Quarter, 1996.

[2]. Personal Interview, Elliott Hamilton, EMCI Consulting, Washington DC, 5 April 1998.

[3]. CDMA Development Group, Press Release, New Orleans, LA, Feb 8, 1999.

[4]. CTIA, "Wireless Factbook", Washington DC, Spring 1995.

[5]. CDMA Development Group, Press Release, New Orleans, LA, Feb 8, 1999.

[6]. Dr. George Calhoun, "Digital Cellular Radio", p.69, Artech House, MA. 1988.

[7]. Stuart F. Crump, Cellular Sales and Marketing, p.2, Creative Communications Inc., Vol. 5, No. 8, Washington DC, August 1991.

[8] Herschel Shosteck, "The Retail Market of Cellular Telephones", Herschel Shosteck Associates, Wheaton, Maryland, USA, 1st Quarter, 1996.

[9]. Telecommunications Industries Association Transition to Digital Symposium, "New Services and Capabilities", TIA, Orlando, FL, Sep 1991.

[10]. Interview, Elliott Hamilton, EMCI, Washington DC, 6 March 1996.

[11]. Stuart F. Crump, Cellular Sales and Marketing, p.1, Creative Communications Inc., Vol. 5, No. 8, Washington DC, August 1991.

[12].ibid.

[13]. Cellular Integration Magazine, "Techniques", Argus Business , January 1996.

[14]. Interview with Steve Kellogg, APDG Research, June 3rd, 1998.

[15]. United States Patent 3,663,762, "Mobile Communication System" Assigned to ATT, May 1992.

[16]. Letter to the TIA voting members from Eric J. Schimmel, 21 November, 1990.

[17]. Personal interview, "Jeffrey Schlesinger", UBS Securities, New York, February 12, 1996.

[18]. Personal interview, Bob Glen, Sparton Electronics, Raleigh NC, January 1996.

[19]. Personal interview, "Jeffrey Schlesinger", UBS Securities, New York, February 12, 1996.

[20]. Personal interview, industry expert, July 15th, 1998.

[21]. Personal interview, industry expert, July 15th, 1998.

[22]. D.M. Balston, R.V. Macario, "Cellular Radio Systems", Artech House, 1993, pg. 223.

[23]. "Double the Coverage or Half The Cost", Reflection Technologies, April 1999, Sunnyvale, CA USA.

[24]. Personal Interview, Elliott Hamilton, EMCI Consulting, Washington DC, 25 February 1996.

Chapter 8
Mobile Phone Features and Services

The features available in CDMA mobile telephones may vary from manufacturer but generally are combined with the CDMA basic services to produce advanced services. Some features can only be performed in the mobile phone (e.g., multi-language menu), other features can only be implemented in the system (e.g. call conferencing) and some features can be implemented in either the mobile station or system.

This chapter describes mobile phone features that are available now or will be available in the next few years. The CDMA industry standards are continually evolving to allow new features. Because some advanced features require communication with the CDMA network, some features (such as data services) systems have not been offered by service providers.

Using advanced messaging features, some CDMA digital phones are able to receive important information like e-mail, news, weather reports, sports scores, stock quotes; and have the ability to conduct on-line transactions such as banking and shopping (e-commerce). The World Wide Web and digital wireless communications are melding together to present the ultimate in everyday lifestyle management and decision-making. This means communications, information, and transactions shall be available at a person's fingertips in one CDMA mobile device.

Mandatory Mobile Station Features

There are features that must be included in a mobile station to allow the operation of basic telecommunication services. These mandatory features include a key-pad or control device, display or status monitoring, and audio controls. There are some optional features such as an accessory connector or extra buttons for menu features or abbreviated dialing, which may vary from manufacturer as well as phone model. Regardless if a feature is mandatory or optional, it must conform to CDMA specifications as defined by IS-95.

Caller ID

Caller ID allows the phone to display the calling number (or name) prior to answering the call. This allows telephone customers to determine if they want to answer the call. Some CDMA telephones have the ability to use the displayed calling number to look-up a name in the telephone's phonebook memory, thus a name can appear in addition the telephone number on the display. .

Short Message Indicator

The short message indicator feature provides the user with a visual and/or audio indication to the user that a short text message is waiting. Generally, the short message indicator is an icon that is a sealed envelope, however, some phone models simply display text that there is a "MESSAGE WAITING". Figure 8.1 shows a common short message indicator. In this example, the short message that is stored in memory is indicated by a flashing envelope on the display.

In the future, it will be possible to reply to short messages directly from the mobile phone. In some wireless systems, the reply feature can be accomplished by either pre-programmed messages (e.g. press 2 for "I'll be Late"), via the keypad or by an external keypad.

Normally the message has already been sent and stored in the phone's memory. The user selects the menu option to read the message from memory. Depending on the wireless system capabilities, it may be possible for the user to request additional information on the status of the message by further interrogation of the service center. The service center can then transmit additional message information to the user.

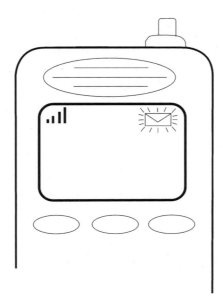

Figure 8.1, Short Message Indictor

Short Message Memory Full

Short messages are typically stored in the memory of the phone. An indication will be given to the user of the phone when the short message memory has been filled. When the memory is full, incoming messages cannot be received. The Short Messaging Service Center (SMSC) treats these messages as if the mobile unit were turned off, and stores the messages until the maximum SMSC storage time has elapsed; at which time the messages are deleted from the SMSC. To begin receiving messages again, the user must empty part or all of the short message area.

Data Accessory Status Indicator

When a phone is connected to a data terminal equipment (DTE) device, the phone should display the specific accessory being used in conjunction with data services. This allows the customer to know the status of the phone (data mode) and the device that is connected to it (ready). Some examples of DTEs are PC modem cards and fax machines.

Indication of Call Progress Signals

Call progress signals are used to provide the user with an indication of the status of a call. For example, if the system is busy, the user will hear a tone and the display may say "System Busy."

Battery Level Indicator

To allow the user to know how much battery life is left in their phone, a battery capacity indication is provided. The battery level indicator may be in the form of a number of graphic bars or a numeric indicator that decreases as the phone's battery is discharged. Additionally, an audible tone or beep may be sounded when the battery is low and/ or a text message, "Battery Low", may appear on the display.

System Indication (ROAM)

The system indication provides the user with the ability to determine whether they are "roaming" or not. The network in use may be indicated on the display.

Over the Air Phone Number Programming

CDMA gives the carrier complete management/control with over the air programming (called intelligent roaming) of various aspects of the phone. This includes the down loading of the MIN (mobile identification number), displays/messages, memory addresses, system selection - download of the Preferred Roaming List (PRL).

Preferred System Selection

When more than one CDMA network is available in a given area, the user will be given the option for selecting a preferred system. Additionally a preferred system list may be stored in the phone as defined by your carrier.

Keypad

A keypad or other control device provides the physical means of dialing numbers. A typical keypad layout is shown in figure 8.2. The actual number of keys, and keypad layout may vary by manufacturer and model. Additional dedicated keys may provide the means to control the Mobile Station (e.g., to initiate and terminate calls).

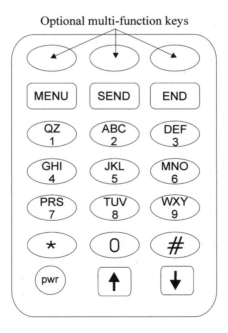

Figure 8.2, Typical CDMA Keypad Layout

Because many new features are available for CDMA handsets, the number of keypad buttons has become a challenge. The typical CDMA phone has 16 to 21 buttons. Some of these buttons may be on the side (such as the volume control) additionally phones are containing "soft" keys whose functionality changes depending on the feature selection. Also the use of a mouse-like or track ball type key sometimes called a navigation key (or "Navi-key") is becoming popular. The mouse key is a four-way rocker key that gives mouse-like navigation for ease of menu and phone book use. The trend is to simplify the user interface, and keep the keypad uncomplicated. Even with 21 buttons, there are typically many more features than buttons. This requires buttons to have multiple uses. For example, the volume buttons may also be used to control the menu operation. Some typical buttons and their use include:

> Digits 0 through 9 are alpha entry- call dialing and control of telephony features such as voice mail
> * and # - used to control some telephony features such as voice mail control
> Power - on/off function for the phone
> Menu/Function - used to activate menu feature on display
> Send/Talk - used to initiate calls
> End - used to finish calls
> Up/Down or +/— usually has combined use for control of the audio volume, menu selection, and feature level such as ring volume.
> Clear - used to clear an entry on the display or to reset a feature selected via the menu

Radio Signal Indicator

The radio signal indicator provides the user with an indication of the approximate amount of radio signal strength available to initiate and receive calls. This is also called the radio signal strength indicator (RSSI). The radio signal indicator is usually composed of multiple bars. As the signal becomes stronger, the number of bars that is displayed will be increased.

Emergency Call Capabilities

The phone has the capability to store "emergency" call numbers. This number can be stored in the phone memory. The number, generally "911" in the USA, can vary by location depending on services offered by the municipality or state. It may also in other countries, vary,. To activate an emergency call, a key sequence may be used or for some phone models, the user could simply press and hold any of the buttons (e.g. "9" key) for an extended few seconds.

Dual Tone Multi Frequency (DTMF)

Dual-Tone Multi Frequency (DTMF) tones are often used to control phone devices such as interactive voice response (IVR) systems or voice mailboxes. CDMA phones are capable of allowing or disabling the sending of DTMF tones after a call has been initiated. One of the reasons for the capability to disable the DTMF tones, during a conversation, is to allow keypad commands to be entered in a such a way that the storage of a phone number into memory doesn't create the DTMF tones, which may annoy the listeners.

Repeat Dialing

Mobile telephones may be able to automatically repeat their request for service (repeat dialing) for a variety of reasons. These include: the called number is busy, no answer or the system is busy. The CDMA network places limits on the number of attempts and intervals (from 5 seconds to 3 minutes) for automatic repeat service requests to limit congestion on the CDMA system. If the mobile phone makes the maximum number of retry attempts, the phone will automatically stop requesting service until the user re-enters the request via the handset. Additionally, a phone may offer a user programmable automatic redial function depending on manufacturer and model.

Optional Mobile Station Features

There are many optional features that may be performed by the mobile station. These features are what help to differentiate one manufacturer's product from another.

Direct International Calling

Direct international calling allows the mobile phone user to have a standard method of initiating international calls regardless of the system or country they are dialing from. For this purpose the phone may have a key whose primary or secondary function is marked "+". This is signaled over the air interface and would have the effect of generating the international access code in the network. It may be used directly when setting up a call, or entered into the memory for abbreviated dialing. This feature is of benefit since the international access code varies between countries, which might cause confusion to a user, and prevent the effective use of abbreviated dialing when roaming internationally. Users may still place international calls conventionally, using the appropriate international access code.

Call Restriction

There are three types of call restriction that can be enabled in the mobile station: fixed number, outgoing, and incoming. The restriction may be user selective, such as its applications to telephony or data transmission service or call types such as long-distance or international calls. These restrictions may also be controlled by the network, as CDMA provides the carrier with complete control/ management of many of the features & functions of the phone via "over the air" activation & programming (e.g. network does not permit international calling; or blocking calls to specific countries. Some of these carrier imposed limitations are for security, anti-fraud measures).

Fixed Number Dialing

Fixed number call restriction only allows the user to dial numbers that have been preprogrammed into the phone memory. To dial numbers not stored in the phone memory, an electronic un-lock code is required. Dialing restriction should not affect the ability to make Emergency Calls (for example "911").

Incoming Call Restriction

Incoming call restriction disables the mobile phone from answering incoming calls. This feature would be used on mobile phones that offer pre-paid telephone service for customers who want a mobile phone for placing calls only. Because some countries bill the owner of the mobile phone for incoming call usage, it would be disadvantageous to the network service provider if the pre-paid user discovered the phone number of the pre-paid telephone and used the phone for incoming calls (for which the user would not be billed). With incoming call restriction, even if the customer discovers the phone number of the pre-paid phone, calls cannot be received to the phone.

Outgoing Call Restriction

Outgoing call restriction disables some or all of the ability of the mobile phone from originating calls. The outgoing call restrictions may also be user programmable depending on manufacturer & model. These include local calls only, single number calls, pre-programmed numbers, and domestic calls only (no international calls).

Local only calls typically restrict the phone to a fixed number of dialed digits. For many countries, this may be 7 digits for local and 10 digits for long distance. Because of the rapid growth in telephone number assignments, some local telephone regions have been divided and they require additional digits for local dialing. To enable local call restriction to these areas, the mobile phone may be allowed to access only specific regional codes (area codes).

Single number calls often restrict the dialing to a single number. This single number might be a company office number of a call center (toll center) that allows the redirection of the call to another number. If the phone was forced to dial the single number that dials the call center, the call center could track the usage of all calls originated from the phone.

Some phones only allow the dialing of speed dialing numbers. This may be used by a company to restrict calls to the office, shipping department or another phone number that would be used by its employees.

The restriction of international calls may be performed by limiting the maximum number of dialed digits or restricting calls that have the international access code as part of the first few digits.

Power Control and Automatic Registration

Power control allows the power of the mobile station to be turned on and off. The power switch on mobile phones is controlled by a software program that allows the phone to perform specialized operations when it is turned on and off. These specialized operations include notifying the system when it is first turned on and when it is turned off. This allows the system to better track the location and status of the phone. The system can control the power level of phone independent of user control.

Extended Address (sub-address)

This feature allows a mobile to send or receive additional digits when dialing a telephone number (directory number). These additional digits may be used to set-up calls to extensions or other services that use the extra digits.

Enhanced Voice Quality

CDMA utilizes a dynamic rate voice coder that offers high quality voice. This enhanced voice is a new form of digital compression used in CDMA networks. This provides for a rate that is adjusted in a range from 8 kbps to 13 kbps. To use enhanced voice service, the new digital compression technology must be located in the system and in the handset.

Voice Privacy

Encoded transmissions provide greater call privacy and fraud protection. CDMA digital networks provide one of the highest degrees of security in the wireless industry. Eavesdropping on conversations over a CDMA digital network is virtually impossible becaus the signal is diffused throughout the spectrum and encoded at multiple levels.

Voice privacy is a feature offered by some telecommunications systems to prevent the listening of communications by unauthorized users. Voice privacy typically involves the encrypting of the voice signal with a shared secret key so only authorized users with the correct key and decryption program can listen to the communication information.

Battery Saving (Cell Broadcast DRx)

This feature enables a mobile station to save on battery utilization, by allowing the mobile station to not listen during the broadcast of messages the subscriber is not interested in.

Data Transmission Interface

Data transmission interface provides the ability to connect a computer or fax machine to the CDMA telephone. To send data, CDMA phones need a connector that allows the attachment of an external data card interface adapter. This is not a modem. The data interface only adapts the digital information from the device to the digital format needed by the CDMA telephone. Figure 8.3 shows the basic types of data interface accessories.

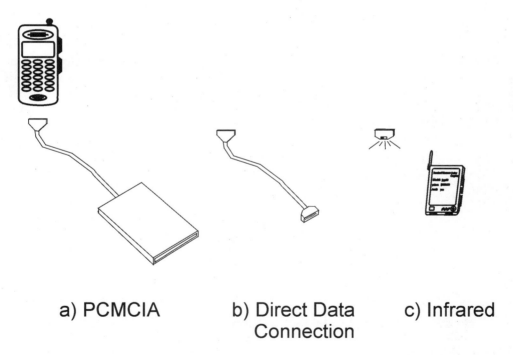

a) PCMCIA   b) Direct Data   c) Infrared
                      Connection

Figure 8.3, Data Interfaces

Service Provider Name

When the phone is in an idle mode, and the network service provider's name is available from the network, the phone indicates the provider's name on the LCD, and stores it in the phone's memory . The capability of displaying the service providers name is also depending on if the network is sending this information.

Broadcast Extended Short Message

The network has the capabilities to transmit various messages to the phone which may appear on the display. This feature allows a mobile station, by support of the extended short message cell broadcast channel, to enhance the capacity of the service. The support of the extended channel has low priority, i.e., the MS can interrupt the reading of this channel if idle mode procedures have to be executed.

Any Key Answer

The any key answer option allows the customer to answer the phone by pressing any key. This key is especially useful in a car which has hands free capability. The user can simply reach down and press any key to answer the call without taking their eyes from the road ahead of them. Normally, any key answer will not use power key and some models of mobile stations do not use "0" key for the "any key answer" feature.

Automatic Answer

After a fixed number of ring signals (typically 2 or 3), the phone may automatically answer. This allows the user true hands free capability when in the car.

Mute Control

The mute option allows the user to select mute for the audio signal. This allows a call to continue without the caller hearing the conversation of the user or background noise.

DTMF Tones

Provision has been made to enter DTMF digits or touch tones, together with a telephone number. When the called party answers, the ME will send the DTMF digits automatically to the network after a delay of 3 seconds (± 20%). The first occurrence of the "DTMF Control Digits Separator" would be used by the ME to distinguish between the addressing digits (i.e. the phone number) and the DTMF digits. Upon subsequent occurrences of the separator, the ME would pause again for 3 seconds (± 20 %) before sending any further DTMF digits.

Automatic Updating of Directory Numbers

The CDMA operators via over the air activation/ programming can down load phone address lists directly into the phone memory using the short message service. This saves the subscriber time in directly programming the phone numbers and name tags into each memory location in the phone.

Short messages can be of various types: Point to Point, Mobile Terminated (MT), or Cell Broadcast. They may be used to convey a directory number which the user may wish to call. This can be indicated by enclosing the directory number in a pair of inverted commas (" ").

If the displayed message contains these characters enclosing a directory number, a call can be set up by user. Normal (unspecified) or International format (using + symbol) may be used. The message may contain more than one directory number, in which case the user must select the one required.

Last Numbers Dialed (LND)

The mobile station may store a pre-specified quantity of most recently dialed numbers (Last Number Dialed, or, LND). Up to ten LND numbers may be stored in the phone's memory; the quantity of LNDs that may be stored in the ME is not specified.

Service Dialing Numbers

Service dialing numbers are pre-stored telephone numbers that allow the customer to easily dial for specific services (e.g., customer care). The main advantage of service dialing numbers is that they are programmable by the system operator. This allows the service numbers to be updated without contacting the customer.

Telephone Lock

Telephone lock allows the user to create and assign a lock code that restricts access to the mobile telephone. The lock mode may be enabled when the phone is not in use. To lock the phone, the user typically enters a key sequence (e.g., Func + 1) and the phone may display the word "Locked." Many phones use a 3 or 4 digit lock code. To help the customer remember the lock code (in case they accidentally lock the phone), the dealer normally programs the lock code with the last few digits of the mobile telephone number.

NAM Programming Lock Code

This feature requires a mobile phone to use a lock key code (different than basic lock restrictions) that has a new number assignment module (NAM) system information programmed into the memory of the phone. This information may include one or more of the following: mobile identification number (MIN), System Identification (SID) and priority access. If a SIM card is inserted that does not match the stored information, the phone will be inoperative.

The purpose of the NAM programming lock code is to ensure the customer must go back to their existing service provider to get the lock code before changing to a new service provider. The NAM lock code is typically unknown to the customer.

Advanced Mobile Station Features

Advanced mobile phone features provide a way to differentiate a manufacturer's mobile station from competing models. Advanced features may vary from manufacturers and by models. The recent CDMA service feature and technology advancements have also allowed for new advanced mobile station features.

Figure 8.4, Graphics Display on Mobile Phone
Source Siemens

Color or Graphics Display

Color displays allow a quick way to differentiate features or services. This would be especially useful when transmitting data via the Internet with links in color. As more complicated graphics become transmitted, color becomes a key element in mirroring computer-like services. Figure 8.4 shows a mobile radio with a color graphics display.

Multi-Language Menu

Many CDMA handsets allow the user to select menu systems that are in different languages, primarily English, Spanish and French. Some handsets are incorporating Chinese, Korean characters as well. With icon-driven displays becoming more popular however, an "all icon" menu or "no-lingual" driven menu is more universally appealing. Figure 8.5 shows a telephone that has several menu languages.

Figure 8.5, Multi-Language Phone (model 5170)
Source: Nokia

Voice Dialing

Voice recognition software, which enables users to speak commands into their phone, allows users to dial phone numbers, access voice mail and other menu items. The spoken "voice" is the most simple and natural man-machine interface. As Voice Recognition algorithms improve dramatically, this interface to the various phone features and functions becomes simpler, more friendly and intuitive. Additionally regulatory and safety issues may help to drive the industry towards this technology's adaptation into mobile usage, as many states in US and several countries have passed legislation to prohibit a driver from using a cellular phone while operating a vehicle. This forces hands free kit or voice dialing. Voice recognition technology may be separated into two categories: Speaker Dependent and Speaker Independent.

Speaker Dependent technology requires the user to "train the speech recognition program. This training requires that the speaker utter the words or names several times, and it's key advantage, is that the user can "personalize" the memory locations with the person's specific name rather than using a generic term such as "home", or "office". The drawback however is that the phone will be limited for the usage only to the individual's voice that the speech recognition program was trained to.

Speaker Independent recognition, which does not require training, permits a broad range of users (provided that they speak in the language of the system, e.g. "English"), has a predefined vocabulary which the user needs to be aware of. Speaker Independent technology does not permit "personalization" to the extent of the speaker dependent system.

Voice recognition software can be resident in the phone or network. If integrated into the phone, voice recognition software is usually speaker dependent for person-

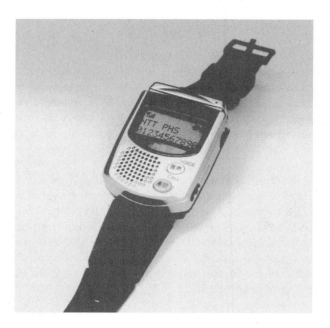

Figure 8.6, Wrist Watch Phone with Voice Dialing
Source: NTT

alizing the memory locations and dialing. Speaker dependent also requires less processing power, so its more economical (than speaker independent) to deploy into the handset. Network operators can utilize either technology or a hybrid of both.

Currently voice recognition software is being integrated into various manufacturer's handsets as well as being deployed by the carriers on their networks. The carriers are viewing network based voice dialing as a value added service, and revenue generating opportunity by charging a monthly service fee.

New voice service & applications have been created by various companies, including Wildfire Communications and General Magic. These new services are "Virtual Assistants" that are very applicable for use in a "mobile" environment. They perform various commands by voice such as dialing, accessing voice mailbox, reading email, getting stock quotes, maintaining a calendar, access to the Internet, etc. The use of such services is of great value to the individual user because these services are personalized to the individual users needs.

Answering Machine

Recording messages is a very popular feature in landline devices. While some systems simply forward to voicemail services which then records the message, the telephone's ability to store short voice messages in its own internal memory saves the user from having to make a call to their voicemail to retrieve messages. The answering machine message in the mobile unit could say "I'm at home right now, please leave a message and I will return your call when I go out again."

Figure 8.7 shows a functional diagram of a phone that has a built in answering machine. In this diagram, the mobile station detects an incoming call and decides to answer the call. The first step involves automatically answering the call (mobile station sends a "SEND") message to the system. The phone then connects the audio to a previously recorded announcement message. After the announcement message has been played, the audio is connected to a storage device (typically digital memory) to store the audio. The audio message is then replayed from the storage device at a later time to the user of the phone.

Figure 8.7, Phone with Answering Machine
Source: Motorola

Digital Answering Machine

The phone automatically answers, plays an audio message requesting that the caller to dial in the phone number to return a call. These digits are stored in memory and can be automatically dialed by the user at a later time.

Voice Memory

This feature mimics the benefits of voice digital recorders such as "VoiceIt", whereby users can record reminders to themselves and play them back later. Another useful advantage of this type of feature on a portable phone is the ability of the user to record a phone conversation in which directions were given or a phone number dictated, or anything that the user might want to remember. This feature becomes especially useful if it has the ability to sort and scroll quickly through messages. Due to the limited memory that is available in a cellular phone, the recording time maybe very limited - i.e. several 15 or 30 seconds slots to record a message.

Email Access

CDMA networks have messaging capability to send short messages directly to a mobile telephone, many CDMA operators have set up messaging centers that transmit email messages directly to the phone. When using short messaging to receive emails, the message must be in the correct format. There are several web sites that allow direct entry of an email message that is to be sent to a CDMA mobile telephone. When using short message service to receive emails, the limitation on message length is 160 characters per message.

To display a 160 character message on a small telephone display, the message typically scrolls across the display. Short message service stores messages into the phone memory. Limitations on the memory of the phone limit the maximum amount of email messages that can be received. Another option is to receive email messages using data transmission service to a laptop computer or other portable computer devices that have email capability. The SMS feature can provide an e-mail notification that indicates to the user the beginning of a download of a longer e-mail data message that may be transferred to a PC.

Information Services (Web Browsing)

CDMA provides a robust platform for data handling and transmission. Depending on the CDMA operator, services such as text messaging and Internet-related information services such as headline news, sports and weather may be offered. In the longer term, CDMA's expansive data capabilities will support interactive transactions and transfers of larger volumes of data.

Web browsing involves the ability to interact with Internet web sites via the CDMA network. Browsing the web from a home or business computer via the public telephone network ordinarily involves full keyboard control, graphics display and moderate speed data (28 kbps or more). Most CDMA telephones have a limited number of keys, a small display without graphics capability and relatively low data transfer rates (9.6 kbps).

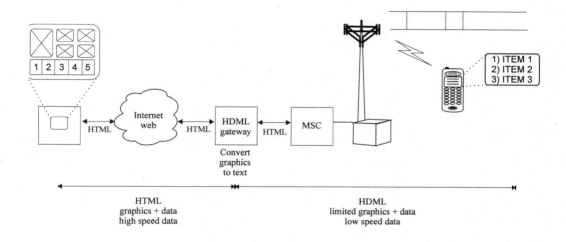

Figure 8.8, Connecting to the Web Using HDML

To access the internet via a CDMA telephone, much of the graphics must be removed due to their large memory size. There are several alternatives to remove the graphics. One of the more popular alternatives, in use at this time, is the Handheld Device Markup Language (HDML). The HDML system was developed by the company Unwired Planet Inc. HDML is very similar to HTML. The primary difference is HDML is optimized for devices that have low data transmission rates and small screens. More information about Unwired Planet and HDML can be found at the Unwired Planet website: www.uplanet.com.

Figure 8.8 shows how an HDML system works with a CDMA phone. In this example, a CDMA mobile phone is communicating with a web site through an HDML gateway. This gateway converts the high resolution graphics and mouse controlled options to text based items that can be controlled by the phone. Also handset must have WDML "microbrowser" in the phone.

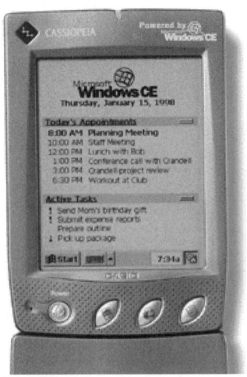

Figure 8.9, PDA with Email Capability
Source: Casio

Accessory Display and Control Device

A device that adds more keypad controls (such as a keypad) and larger display (such as a tablet display) allows more advanced services to be accessed via the telephone. This device can be linked to the handset by a wire, infrared device (supported by windows CE) or by radio. Portability becomes a concern and the question arises in how many ancillary devices a user is willing to transport. A wireless device built into a laptop computer may be more appealing to end-users.

Ring Options

CDMA phones have many different ring options: incoming call, short message received, and alternate phone number ring (for a 2nd line). Ring options allow the user to change the sound of the alert tone. To set the ring options, the user enters the menu mode and selects the alert or ring mode to be modified by using the up/down keys. Ring options dramatically vary from manufacturer and by model.

Mechanical Alert

If available, an incoming call announcement may be provided by the use of a mechanical vibrator. This vibrator can be built into the phone, into the battery, or supplied as an accessory that clips on to the bottom of the telephone. Similar to the setting of ring alert preferences, it may be possible to set different vibrate modes such as short vibe bursts and slow to fast vibration settings for different types of alerts.

Phone Book

A phone book can typically hold telephone numbers and the names associated with the telephone numbers (called name tags). Some CDMA phones allow the automatic display of the name tag when the calling number identification matches the corresponding number stored in memory.

To store and access letters into the phone, most phones use the letters on the keypad. Because each key normally has 3 letters associated with it, the user must press the key multiple times to enter the letter. After the correct letter has been selected, the user may press an up/down key or wait a short time until the next position become highlighted. This indicates it is ready for another character.

Each phone number is given a number location in memory. Because it is unlikely that the user will remember the number, search capabilities exist to allow the user to find the name by using letters. To search by name, the user enters a search mode (possibly by a menu option) and uses the letter keys in a similar manner used to store the names. As the letters are selected, the closest name match is displayed. The user then is allowed to scroll through all the names in alphabetical order from that point using the up/down keys.

Call Timers and Logs

Call timers can be used by consumers or businesses to review the approximate amount of airtime usage. Some phones have call logs that list telephone numbers as well as the duration of calls. This is a valuable feature for rental and phone leasing companies. It is also helpful for corporations that track the usage of each employee.

Speed Dialing

Speed dialing allows a user to dial pre-stored telephone easily and fast. Speed dialing is made possible by storing telephone numbers into memory and allowing the user to access those stored telephone numbers through the selection of a limited number of keys (usually one or two digits). Newer speed dialing features include access to the last 10 numbers dialed, and the sorting of speed dial numbers in different folders.

Software Updates

CDMA phones have the ability to transfer new software into the telephone via over the air activation programming. Software updates can be sold or given gratis to provide fixes to software bugs or to add new features to the mobile station. Some digital telephones do have the capability of updating a portion of their software parameters through messages sent via the cellular or PCS network. This facilitates undetected improvements and circumvents most system down times, a feature not recognized, but certainly appreciated by end-users.

Personality Identification

Personality identification bonds the user to their phone or service. Personalization includes a wake-up message, such as "Good Morning Bob", customization of features, such as auto-storing of numbers in pre-defined memory locations, scheduling information and calendar functions as well as personal account information, calling card information and other useful information/reminders. As product development embraces "lifestyle marketing" this feature is likely to become the cornerstone in these marketing efforts.

Clock

A clock can be included with the phone that maintains the time. The CDMA network has the capability to automatically update the time zone where the phone is operating. This can automatically update the clock time. Having a clock requires the inclusion of a battery to maintain the clock time when the battery of the phone is removed or discharged.

Time Delay Dialing

Time delay dialing allows the user to program the time delay prior to calling. This feature allows the user to schedule a retry attempt (e.g., 20 minutes). Prior to initiating the call, the mobile phone will notify the user it is dialing by an audio sound.

Theft Alarm

The theft alarm monitors unauthorized usage of the mobile phone in the event it is in the locked mode. If the phone is tampered with (e.g., repeated incorrect attempts to unlock), the phone will automatically dial a preprogrammed number (e.g., alarm monitoring station).

Hands-Free Options

Hands free operation is the use of a mobile station without the requirement of holding the handset while talking. The hands free option may be a simple solution such as a portable microphone and ear piece to a car mounted speaker and directional microphone. This feature is especially useful while driving, or during a meeting when notes need to be taken. While the quality of hand free systems has been less than ideal in the past, it is improving. Some units have a motion sensor that allows the muting or disabling of hands free when placed to the head or when a hand is waved in front of the handset.

The ability to plug a handset into a portable hands free speaker is sometimes called a "Poor Mans Hands-Free." Caution should be exercised however as the ear piece can cause deafness if volumes exceed certain levels.

In mobile stations with group call, an internal high volume speaker may be included to allow the continuous monitoring of group call or audio broadcasting while the phone is near the user (such as clipped on a belt). When the loud speaker is in use, the volume is much higher than standard volume. This could result in injury to the user when held close to the ear. To overcome this potential challenge, phones that have a high volume loud speaker may include a proximity detector. The proximity detector senses the closeness of the phone to other objects. When the proximity detector determines it is close to an object (such as an ear), it can automatically reduce the volume of the speaker

Soft Keys

Soft keys are buttons on the keypad of an electronic device that have the ability to redefine their functions. Soft keys are typically located adjacent to a display that provides a description of the key function. This allows an electronic assembly to reduce the number of keys that are required to operate all of its features. This is especially important for portable handheld telephones where the maximum number of keys are limited by the size of the device.

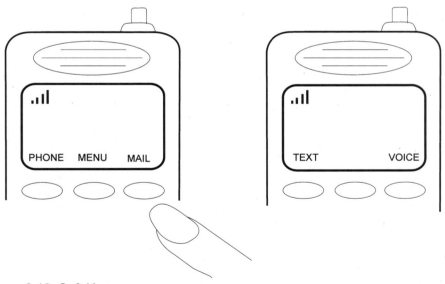

Figure 8.10, Soft Keys

Figure 8.10 shows a sample of soft keys. In this diagram, a user selects the option to look into their mailbox using a soft key. After they have selected the mailbox function, the soft key is redefined with a new function (voice message).

Electronic Help Instructions

Electronic help instructions contains the operator's manual that is stored in the memory. The key to successful electronic help instructions are the menu access categories.

Macro Function (Multi-Key Sequence Programming).

Macro functions allow the user to store a sequence of operations that will be used later through a reduced number of keypad operations. For example, the key sequence Function + 4 may dial the users voice mail system, wait a few seconds for an answer, then dial a mailbox number and then dial an mailbox access code.

Calling Card Dialing

Some phones come specifically equipped for calling card dialing. This is a specific macro function that allows the use of calling cards. To initiate calls via a calling card, this ordinarily involves three different numbers: a calling card telephone access number, the dialed digits and the calling card access code. Because it may take some time for calling card systems to connect calls, the calling card dialing feature usually allows the user to insert pause time periods between dialing numbers. The final result of the calling card dialing feature is a procedure that allows the user to enter the desired number and press the calling card operation. This may be a specific key, function key sequence, or a soft key. The phone then dials all the necessary numbers.

Fixed Wireless Phone

This phone would offer all the same features of a landline phone, e.g. call transfer, call waiting, conference calling, etc but doesn't require installation. This unit can be plugged into any electrical outlet. The unit has two watt transmit power which will provide for greater call coverage.

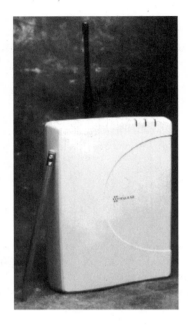

Figure 8.11, Fixed Wireless Phone Interface
Source: Telular

VIDEO/ Picture Telephone

A picture telephone combines a miniature camera and a graphics-capable display into a mobile telephone. Picture phones allow emergency workers, police or insurance adjusters to send and store pictures to achieve real-time assistance.

Chapter 9
System Features and Services

System features and services are typically provided by call processing software in the CDMA network. Using any mobile station can operate some system features (e.g. call forwarding), some features require interaction with software contained in the mobile station (e.g. short message service) and some system features require any interaction with the mobile station (e.g. call restrictions).

The CDMA system is capable of providing many types of voice, messaging and data services[1]. Most of these services are compatible with services offered by other types of networks (typically the public telephone network). To ensure these services can correctly operate with each other, the CDMA specifications detail many of the precise operational parts of these services and how they are expected to interact with other services. There are two basic types of services: bearer services and teleservices. In addition to these basic groups of services, there are supplementary services that combine multiple bearer and teleservices to create advanced services (such as call forwarding) for customers. The introduction of features and services in the CDMA network are released in phases. The CDMA standards continue to evolve to offer the features and services required by the marketplace, which may also vary by carrier. Carriers may charge additional fees for some of the network based features and services and some of the services that are offered depend on the carrier.

---

[1]. During 1998, carriers were NOT offering digital data transmission on CDMA carrier channels. Data transmission for CDMA systems is initially being deployed on a per carrier basis (current data transmission is in the analog mode).

Bearer Services

Bearer services are telecommunication services that are used to transfer user data and control signals between two pieces of equipment. Bearer services are typically categorized by their information transfer characteristics, methods of accessing the service, interworking requirements (to other networks) and other general attributes.

Information characteristics include data transfer rate, direction(s) of data flow, type of data transfer (circuit or packet) and other physical characteristics. The present CDMA system offers two data transfer rate sets. Rate set 1 has a maximum data transfer rate of 9600 bps and rate set 2 has a maximum data transfer rate of 14.4 kbps. Each of these can be divided by half, quarter and one-eighth data transfer rates. The access methods determine what parts of the system control could be affected by the bearer service. Some bearer services must cross different types of networks (e.g. wireless and wired) and the data and control information may need to be adjusted depending on the type of network. Other general attributes might specify a minimum quality level for the service or special conditional procedures such as automatic re-establishment of a bearer service after the service has been disconnected due to interference. The categories of bearer services available on the CDMA system include synchronous and asynchronous data, packet data and alternate speech and data.

Figure 9.1 shows a typical CDMA bearer service. In this diagram, a mobile radio desires to send data to a computer that is connected to a public telephone network (at the office). In this example, the bearer service is circuit switched data. The computer uses a modem to send the information through the public telephone network to the CDMA system. The CDMA mobile telephone calls the CDMA system indicating it wishes to place a data call (access attribute). The CDMA system accepts the call and the mobile phone begins to send data directly on the CDMA radio channel at 9.6 kbps (information attribute). The CDMA system routes this data to a modem that converts the digital signal to a modem signal to be sent to the remote computer (interworking function). The CDMA system uses the phone number sent by the mobile phone to dial the remote computer (general attributes). After the computer answers, the modem signals exchange signal training information and 9.6 kbps data transmission continues (the bearer service).

Teleservices

The IS-95 CDMA specifications call for providing a robust platform for data handling and transmission. As of the end of 1998, carriers only offered only messaging

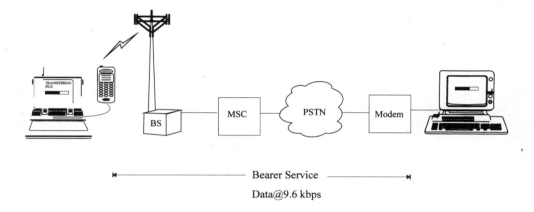

Figure 9.1, Bearer Service

and text based Internet-related information services such as headline news, sports and weather. Longer term, CDMA's expansive data capabilities will support interactive transactions and transfers of larger volumes of data. In the future, mobile phones will have the necessary software to facilitate the cable connection between the mobile telephone and a data device (computer, fax machine) and the network will allow digital data transmission capability.

Teleservices are telecommunication services that provide the user with the necessary capacities including terminal equipment functions, to communicate with any other users. Teleservices are categorized by their high level (application) characteristics, the low level attributes of the bearer service(s) that are used as part of the teleservice and other general attributes. High level attributes include application type (for example voice or messaging) and operation of the application. The low-level description includes a list of the bearer services required to allow the teleservice to operate with their data transfer rate(s) and types. Other general attributes might specify a minimum quality level for the teleservice or other special condition. The categories of teleservices available on the CDMA system include voice (speech), text messaging, facsimile and group voice.

Figure 9.2 shows a typical CDMA teleservice. In this diagram, a mobile user wishes to send a fax from a fax machine that is connected to their mobile phone. After the user dials the destination fax number, the phone detects that a fax machine is connected (general attribute) and requests fax service (teleservice) from the CDMA system. In this example, the bearer service portion of the fax service is circuit switched data with a data rate that is selected by the teleservice and is adjustable up to 9600 bps. The CDMA system routes the fax data received from the mobile phone to a fax modem. The fax modem is used to send the information through the public telephone network to the destination fax machine. As the CDMA mobile telephone sends fax data to the CDMA system, the system checks to determine if the data has been received without error. If the data was received in error, a request

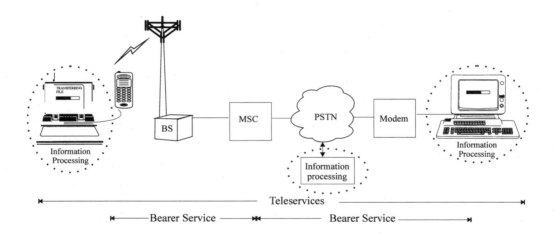

Figure 9.2, Teleservices

will be sent for the mobile phone to retransmit the lost portion of fax data. The fax teleservice ensures a fax can be sent through the CDMA network without any errors due to radio transmission.

Supplementary Services

Supplementary service modifies or enriches basic telecommunication services. A bearer service (e.g. circuit switched data) or teleservice (e.g. voice service) can be a telecommunications service. Supplementary services modify or provide for additional processing of basic teleservices.

Examples of supplementary services include voice mail, call forwarding, call restrictions and others. Some supplementary services can allow interaction with the subscriber (such as call forwarding). Figure 9.3 shows the relationship between bearer service, teleservice, supplementary services and telecommunication services. This example shows that supplementary services can be composed of one or more bearer services and teleservices.

Voice Services

CDMA specifications are in a state of evolution calling for various voice services, however at this time, current systems do not support all of these services. In the future, these services may include voice group call service (VCGS) and voice broadcast service (VBS), point-to-point voice services (handset originated text messages) and group calls.

TEXT Messaging Services

TEXT message service (TMS) gives mobile phone subscribers the ability to receive text messages. TEXT messages are limited to 100 alphanumeric characters which includes spaces. Messages maybe sent from the Internet or through the carriers text messaging service center.

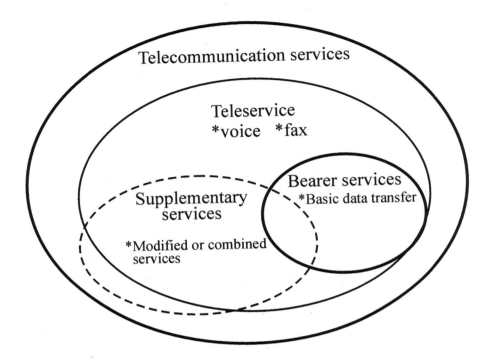

Figure 9.3, Types of Services

Data Services

It is anticipated that CDMA will provide several types of data services that allow the sending and receiving of computer and FAX data. This is an important key to a truly mobile office. Unfortunately, wireless data and FAX signals can have transmission problems. This is why CDMA will support different types of data services.

Circuit Switched Data

Circuit switched data provides for continuous data signals, packet switched data allows for very short messages and fax service allows the verification that complete fax messages are received without errors from radio transmission.

Figure 9.4 shows how the CDMA system will send circuit switched data. In this example, a laptop computer is sending a file to a computer in an office. The laptop computer has an internal CDMA data capable mobile station. The laptop MS first sends a service request indicating the destination phone number along with a request that the call is a data call. When the CDMA MSC receives the request, it connects the call to one of several modems in the mobile switching center. These modems are the inter-working function (IWF) that converts the radio data in to a modem signal that the landline modem can understand. The MSC then dials the telephone number of the destination computer and the modems begin to communicate with each other. The telephone line modem is used to convert the analog data signal back to the original computer data signal that is sent by the laptop computer. After the modems have completed training and adapted their data rates, the data connection between the laptop computer and office computer is complete and the file transfer can begin.

In 1998, the maximum data transfer rate was limited to 14.4 kbps (rate 2). The CDMA specification may allow for high-speed circuit switched data services

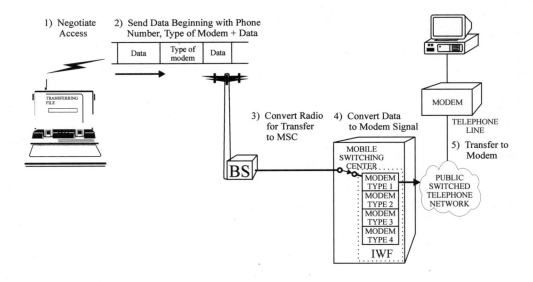

Figure 9.4, Circuit Switched Data Services

(HSCSD) to bring data speeds up to 57.6 kbps or more. CDMA can accomplish this through the combining of coded channels or through the use of a new channel coding processes.

Packet Switched Data

CDMA may provide packet switched data transmission in the near future. The CDMA network was designed primarily to offer voice services. Shortly after the CDMA system was introduced, circuit switched (continuous) data services were developed. The operation requirements for circuit switched and packet switched data services are very different. Circuit switched data has substantial call setup time and is inefficient for serving sensing control and applications that require small amounts of information. The CDMA system must be enhanced to offer packet radio service. Typical applications for packet switched data include Internet browsing, wireless email, train control system, route guidance, credit card processing and many other applications that require small amounts of data when communicating.

Packet switched data transmission provides effective use of the system resources when transferring small amounts of data. Packet switched data service only uses radio resources when there is information to transfer. This provides the advantage of charging only for the amount of information used and increased system efficiency.

The CDMA system is likely to use an enhanced CDMA traffic channel to coordinate packet access. Packet switched data will likely provide for high peak data rates without long connection set-up delays. When data compression is used, user data transfer speeds may exceed 100 kbps.

Fax Services

It is expected that by 1999, fax services will be supported by some CDMA carriers. Fax transmissions carry data that is similar to those of cellular modem transmissions. However, while modem algorithms have been adapted to work well with sometimes poor cellular connections, FAX algorithms have not. Most FAX transmission schemes do not recover or correct errors well. Depending on the error type and location, some or all of a wireless FAX page can be lost, or the FAX page may defaced with black or white streaks as a result of data transmission errors. With newer digital data transmission protocols, this need not be the case. Digital data transmission protocols can correct errors through an encoding scheme and through re-transmission of missing or incorrect data. Assuming there are no dropped calls, wireless FAX transmissions could be as reliable as on a wire line.

Supplementary Services

Supplementary servcies combine bearer and teleservices to provide for advanced features. Some of these services are offered at an additional cost by the carrier.

Call Waiting

Call waiting notifies a CDMA mobile telephone user that another incoming call is waiting to be answered. This is typically provided by a brief tone that is not heard by the other callers. Some CDMA telephones are capable of displaying the incoming phone number of the waiting call.

Figure 9.5 shows how call waiting may be used. In this example, an MS user is talking with a caller from phone 1. A caller from phone 2 initiates a call to the MS telephone number (step 1). When the system senses that a call is in progress and the call-waiting feature is authorized, the system sends a brief audio alert message to the MS (step 2). The user hears the message and presses the "SEND" key to send a flash message to the system indicating the user desires to talk to the incoming caller from phone 2 (step 3). When the system senses the flash request, it places the caller from phone 1 on hold and connects the user to the caller from phone 2 (step 4).

Call Hold

Call hold allows a user to temporarily hold an incoming call, typically to use other features such as transfer or to originate a 3rd party call. During the call hold period, the caller may hear silence or music depending on the network or telephone feature.

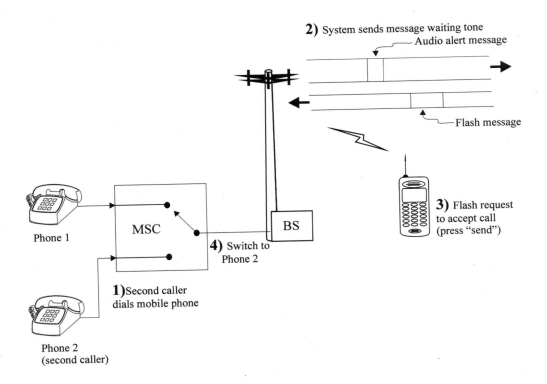

2) System sends message waiting tone
Audio alert message

Flash message

BS

MSC

Phone 1

3) Flash request to accept call (press "send")

4) Switch to Phone 2

1) Second caller dials mobile phone

Phone 2 (second caller)

Figure 9.5, Call Waiting

Figure 9.6 shows the basic call hold process. In this example, the MS user is communicating with a caller from phone 1. The user decides to place the caller on hold and initiates a hold request message by pressing the "SEND" key (step 1). When the system senses the flash message (hold request), it temporarily disconnects the user from caller on phone 1 without disconnecting the call from the MS or phone 1 (step 2). When the call is on hold, the MS can then access advanced features such as call conferencing, call forwarding or other services (step 3).

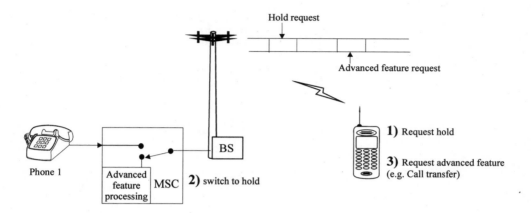

Figure 9.6, Call Hold

Call Forwarding

Call forwarding allows a user to have telephones calls automatically redirected to another telephone number or device (such as a voice mail system). There can be conditional or unconditional reasons for call forwarding. If the user selects that all calls are forwarded to another telephone device (such as a telephone number or voice mailbox), this is unconditional. Conditional reasons for call forwarding include if the user is busy, does not answer or is not reachable (such as when a mobile phone is out of service area).

Figure 9.7 shows two possible ways a call can be forwarded. In the On Demand example, the MS user is first alerted of an incoming call (step 1). When the user sees the calling phone number (caller identification), the user decides to forward the call to another number. The user does this by requesting the call-forwarding feature and dialing the new destination phone number (step 2). When the MSC receives the call forwarding request, it dials the forwarding number (step 3) and connects the call between the caller from phone 1 and phone 2 (step 4).

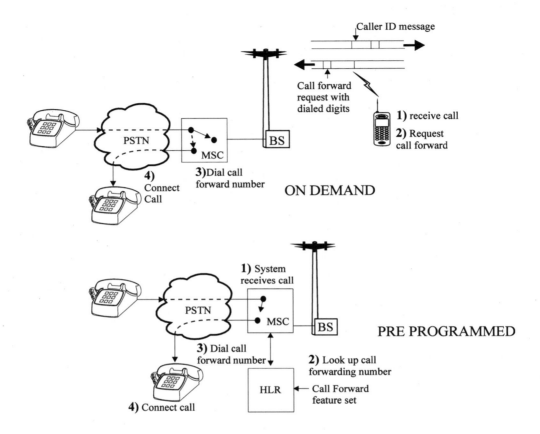

Figure 9.7, Call Forwarding

Figure 9.8 also shows an alternative method of call forwarding which involves pre-programming the destination phone number. In this example, an MS user does not want to be disturbed and they have forwarded all calls to another telephone number. When an incoming call is received from phone 1 (step 1), the system determines the MS user has pre-programmed a forwarding number (step 2) so the system dials the forwarding number (step 3) and connects the call between the caller from phone 1 and phone 2 (step 4).

Call Conferencing (Three Way Calling)

Call conferencing allows two (or more) users to be connected to the same call. Figure 9.8 shows the call conferencing process. In this example, the MS user is talking to a caller on phone 1. The user decides to conference in another person on phone 2. The user initiates this process by sending an advanced service request to the system by pressing the "SEND" key (step 1). The system receives the advanced feature request and places the call to phone 1 on hold (step 2). This MS user then enters the dialed digits of the phone number for the conference call (step 3). The system then dials the phone number of the call for the call conference (step 4) and after the call has been answered, all the parties of the call can be connected together via the MSC (step 5).

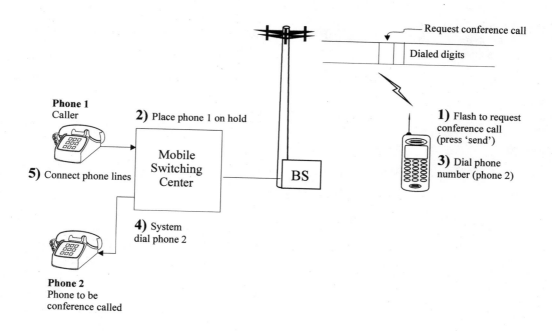

Figure 9.8, Call Conferencing

Calling Line Identification (CLID)

Calling line identification display (CLID) is a service that displays the calling number prior to mobile telephone user answering the call. This allows the telephone customer to determine if they want to answer the call prior to accepting the call.

A caller ID phone number may be transferred in two different ways from a caller to a mobile telephone. Most common is for the phone company to send the calling number to the cellular system. The alternative is for the calling person to enter the digits from a touch tone phone, similar to the method used for sending phone numbers to numeric pagers.

The calling number may be used by the telephone device to look-up a name in memory (e.g. mom) and display the name along with the phone number. Figure 9.9 shows the typical method how the CLID feature can operate. In this diagram, a caller dials the MS telephone number (step 1). The cellular system receives the incoming call along with the digits of the incoming callers telephone number. The MSC decodes the digits from the phone company (step 2) and sends them along with the paging message to the MS (step 3). When the MS receives the page message, it also receives the dialed digits and displays the message on the screen (step 4). Optionally, if the name and phone number is stored internal to the MS, the MS may

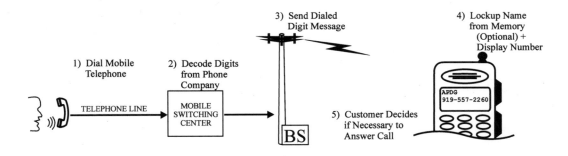

Figure 9.9, Caller Line Identification Display (CLID)

be capable of looking up the phone number and displaying the name that is stored in memory.

The CDMA user is given the ability to block the delivery of their telephone number from being displayed on a caller identification device. This must be requested through the carrier, and availability may vary based on local laws. This feature is called calling line identification restriction (CLIR). CLIR provides the ability for the caller to remain anonymous to the called person. CLIR may be on a per call basis or per line (continuous) block. A message such as PRIVATE or BLOCKED, ANONYMOUS or UNAVAILABLE is displayed to indicate this to the user.

Call Restriction

Call Restriction is a service of prohibiting specific types of calls (e.g. international and long distance calling) from being initiated from a mobile telephone. In addition, it is possible to restrict the delivery of incoming calls (e.g. annoying salesperson) to a mobile telephone.

Voice Mail

Voice mail is a service that allows a subscriber to receive and play back messages from a remote location (such as a PBX telephone or mobile phone). A voice mail system is composed of a storage device (typically a computer hard disk) and an interactive voice response (IVR) control system. When the user dials into the voice mailbox system, the IVR system provides messages that allow the caller to select options to save, retrieve or delete messages.

Network Name and Time Zone

The network identity (name) and time zone feature allows the customer to automatically receive and display the name of the serving CDMA system along with the current local time zone to store and use for clock or timer information. This feature enhances roaming by permitting accurate indication of the CDMA system operator.

Over the Air PROGRAMMING

Over the air activation allows the operator to deliver subscriber information and software updates to the mobile station over the air without the need for "manual" intervention of the subscriber to program the phone. This means that the phone number, features and other subscriber information are automatically programmed using messages from the system rather than laborious keypad input.

Chapter 10

Future Developments for CDMA

CDMA enhancements include intelligent network features, smart antenna systems, satellite systems and uniform communication with information sources and high speed information service access.

Wireless Intelligent Networks (WIN)

To effectively compete against other wireless carriers, CDMA service providers have developed custom service applications for their customers. Wireless Intelligent Networks (WIN) allows a network operator to develop specialized services using advanced intelligent network (AIN) systems. WIN is basically the same as Customized Applications for Mobile Network Enhanced Logic (CAMEL) that is in development for GSM systems.

Some examples of custom services include multiple number service (personal and business) and multiple extension service (one number rings several phones). The CDMA operator or service provider can develop these services on what is called a "services creation node" on the CDMA network.

Figure 10.1 shows an example of a CAMEL application is the creation of an extension phone service feature in a CDMA network. For this example, after a caller dials a subscribers number in a CDMA network, the call is routed to the services creation

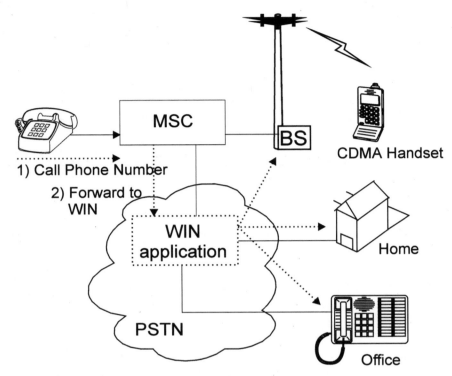

Figure 10.1, Extension Phone Service through WIN Application

node that initiates three new calls; one to the CDMA handset, one to a home tele-phone number and another to an office phone. The first phone to get answered is connected to the calling party.

Wireless Access Protocol (WAP)

A wireless access protocol (WAP) forum was established in 1997 to standardize how digital cellular phones and other wireless devices can utilize Internet content and advanced services. The goals of the WAP forum include the creation of a global wire-less protocol specification that allows applications to operate on wireless devices independent of their type of network access technology (e.g. CDMA, GSM, TDMA, DECT, PHS). The first WAP specifications became available to the public in Feb 1998 and can be viewed at www.WAPforum.org.

The key limitation of wireless information services is the limited amount of data transmission capability offered by mobile systems. WAP defined the boundaries of the Internet and other information application services that have more efficient, simplified or compressed communication requirements. WAP also outlined the significance and uses of a Micro-Browser, similar to the Internet browsing, language scripting similar to JavaScript to provide means for dynamically enhancing mobile device capabilities and WTA/WTAI that allows access to telephone services. WAP also reviews content formats such as calendars (vCalendar) and business cards (vCard).

Figure 10.2 shows the basic WAP structure and its standard interfaces. This diagram shows that multiple types of wireless technologies can be used to interconnect various types of devices for data and messages services. The key part of this system is to standardize the basic parts of the system. The first layer is the wireless transport layer (WTL). This layer is the physical transfer of data and messages. This layer is routed through the wireless transport layer security section (WTLS) to add the authentication and privacy protection to the communication information. The next layer coordinates each communication session. This layer is called the wireless

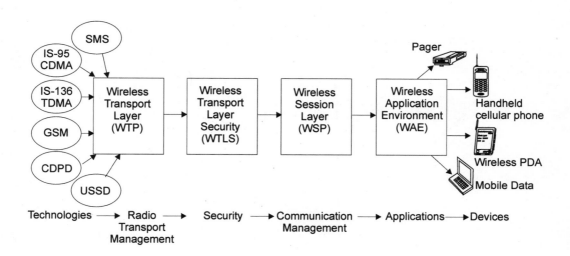

Figure 10.2, Wireless Application Protocol Structure

session layer (WSL). Finally, a wireless application environment (WAE) layer coordinates applications. This layer ensures the application can understand and communicate commands to allow interface to humans or other devices.

Spatial Division Multiple Access (SDMA)

Spatial Division Multiple Access (SDMA) is a technology which increases the quality and capacity of wireless communications systems. SDMA technology increases the system capacity and quality of transmission by focusing the radio signal to narrow transmission beams. Mobile radios that operate outside these narrow beams experience almost no-interference from other radios operating on the same frequency. Because narrow beams focus radio energy, the RF signal can travel further. The use of SDMA allows for cell sites to have larger radio coverage areas with less radiated energy. Because the narrow beam width allows for gain in the receiving direction, this provides greater sensitivity for portable cellular phones which

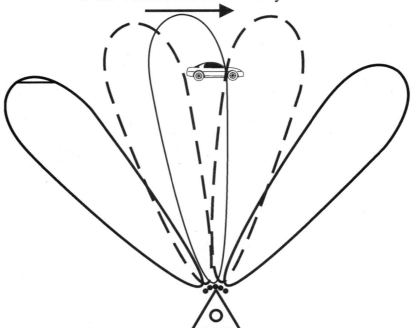

Transfer to New Antenna
as the Vehicle Moves in the System

Figure 10.3, Multi-beam Antenna System

improves the quality of communications channels. There are two basic types of SDMA systems: multibeam antenna systems and smart antennas.

Multi-beam antenna systems use several focused antennas on the top of a cell site (possibly 20 or more) to focus radio energy to specific areas surrounding the cell site. As the mobile radio moves outside each narrow beam coverage area, the cell site automatically transfers the signal to the next adjacent antenna. Figure 10.3 shows a basic multi-beam antenna system.

Using advanced algorithms and adaptive digital signal processing, Antenna systems can be designed to focus their transmission beamwidth dynamically (electrically). This allows them to steer the beams to specific locations of mobile radios. These types of systems are usually called "Smart Antennas." Figure 10.4 shows a smart antenna system.

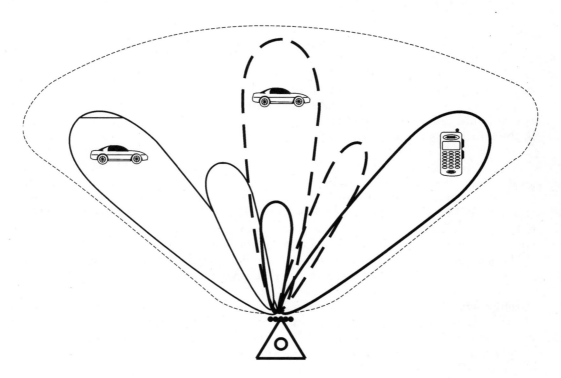

Figure 10.4, Smart Antenna System

One company, ArrayComm, of San Jose California USA, has implemented SDMA into its IntelliCell™ Base Stations. Because of the IntelliCell Base Station's ability to reject interference on the uplink and control transmission patterns on the downlink, Spatial Channels™ can be created. ArrayComm has simultaneously operated three Mobile units on the same radio frequency channel in the same local area. Using an eight element antenna array to track multiple users, the IntelliCell system exhibited Carrier to Interference (C/I) ratios which consistently exceeded 30 dB.

An IntelliCell Base Station that implements Spatial Channels will track the movement of each user in the system. If one user moves too close to another and the users are on the same conventional traffic channel, the system will automatically hand off one of the users to another traffic channel frequency.

Satellite Cellular Systems

Satellite cellular systems replace cell site antennas with antennas that are mounted on satellites. Satellites can provide voice and data services to a wide regional geographic area which terrestrial (land based) services do not reach. These satellite voice systems have been available for many years, however; these systems required large antennas and expensive subscriber communications equipment. There are new satellite cellular systems that have been deployed and planned for deployment in the late 1990's to early 2000's which allow low cost handheld mobile telephones.

Satellite communication systems can provide service to any unobstructed location on the Earth. These satellite systems supplement low cost terrestrial (land based) wireless communications systems [1]. Satellites are much more expensive than terrestrial (land based) cell sites and they have a limited lifespan of 10 to 20 years. To help with the distribution of high costs over many customers, some satellite systems offer frequency reuse similar to cellular technology to serve thousands of subscribers on a limited band of frequencies. These systems use sophisticated antenna systems to focus their energy into small radio coverage areas.

Satellite systems are often characterized by their height above the Earth and type of orbit. Geosynchronous Earth Orbit (GEO) satellites hover at approximately 36,000 km, Medium Earth Orbit (MEO) satellites are positioned at approximately 10,000 km, and Low Earth Orbit Satellites are located approximately 1,000 km above the Earth.

The higher the satellite is located above the Earth, the wider the radio coverage area and this results in a reduction of the number of satellites required to provide service to a geographic area. GEO satellites rotate at the same speed as the Earth allowing the satellite to appear to be stationary over the same location. This allows fixed position antennas (satellite dishes) to be used. Because the GEO satellites hover so far above the Earth, this results in a time delay of approximately 400 msec [2] and increases the amount of power that is required to communicate which limits their viability of hand portable Mobile telephones. GEO systems would include: AMSC, AGRANI, ACeS, APMT [3].

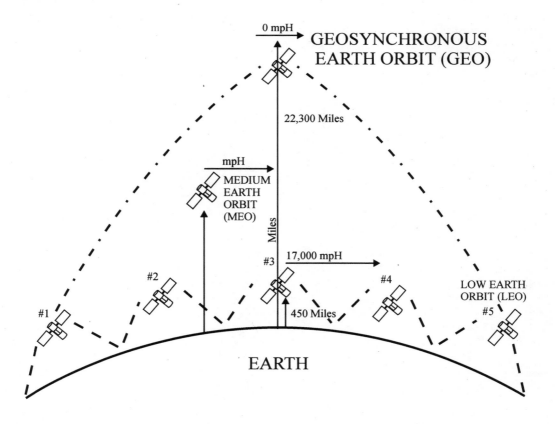

Figure 10.5, Satellite Cellular Systems

The closer distance of the MEO satellite reduces the time delay to approximately 110 msec and allows a lower Mobile telephone transmit power. As a result of the lower position of MEO satellites in the Earth's orbit, they do not travel at the same speed relative to the Earth. This requires the need for several MEO satellites to orbit the Earth to provide continuous coverage. MEO systems include Intermediate Circular Orbit (ICO) Project 21, and TRW's "Odyssey".

Low Earth Orbit (LEO) satellites are located approximately 1,000 km (Iridium 770 km) from the surface of the Earth and this allows the use of low power handheld Mobile telephones. LEO satellites must move quickly to avoid falling into the Earth, therefore; LEO satellites circle the Earth in approximately 100 minutes at 24,000 km per hour. This requires many satellites (e.g., 66 satellites are used for the Iridium system) to provide continuous coverage. The LEO systems include Iridium Inc.'s "Iridium", Constellation Communications Inc.'s "Aries", Loral-Qualcomm's "Globalstar," and Ellipsat's "Ellipso" Figure 10.5 shows the three basic types of satellite systems.

Figure 10.6, GlobalStar Mobile Telephone
Source: Qualcomm

By allowing telephone calls to be routed directly between satellites, some satellite systems communicate directly with each other through microwave links. This eliminates the high cost of routing long distance global calls through other telephone companies. Figure 10.6 shows a picture of a LEO satellite Mobile telephone that uses CDMA technology.

3Rd Generation Wireless

The first generation of cellular technology was analog cellular. The second generation of wireless was digital cellular. Digital cellular technology was developed to replace and enhance the capabilities of incompatible analog cellular standards (AMPS, TACS, NMT, NTT, RC2000, CNET). Although there were fewer digital cellular systems (CDMA, GSM, TDMA), these systems also remained incompatible. The 3rd generation of telecommunications technology proposes a universal mobile telecommunications service (UMTS).

The International Telecommunications Union (ITU) created the concept of the 3rd Generation in 1992 in an effort to consolidate the various 2nd generation standards and combine high speed wired and wireless information services. One of the primary goals of the ITU's UMTS effort is to create one global standard that will be able to allow for a single customer to access fixed or wireless local loop (WLL) networks. This access would use a common air interface and a Family of Systems Concept (FSC). Such a standard is technically possible, although unlikely.

The technical requirements of 3rd generation systems include backward compatibility for 2nd generation handsets and minimum data transmission rates of 144Kbps for mobile applications (cars, etc.), 384 kpbs for portable (walking) and 2 Mbps for fixed applications.

The development of IMT-2000 involves several standards groups including ETSI, ARIB and the TIA. Several industry standards have already been proposed. One of the proposals includes a modified version of the IS-95 CDMA standard — dubbed Wideband CDMA. It is estimated that the ITU will complete an IMT-2000 standard by the year 2000.

IMT-2000 should allow the operation of services that are will beyond the basic radio structure offered by IS-95 CDMA. These services include high-speed data (up to 2 Mbps) and advanced teleservices. Figure 10.7 shows the evolution of the CDMA

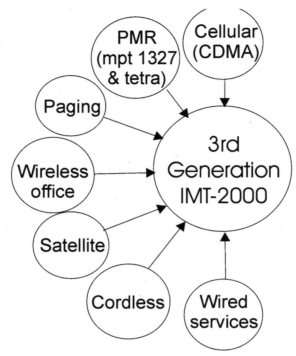

Figure 10.7, Evolution of CDMA Standard

standard towards the 3rd Generation. In this diagram, it can be seen that the vision of the 3rd generation system is the combination of various wired and wireless technologies into a single system.

Advanced intelligent network (AIN) services will be an important part of the 3rd generation system. To allow for new high-speed services that are beyond the capability of the standard IS-95 CDMA radio channel, new access technologies are being developed. There may be a possibility that multiple air interfaces such as wideband CDMA and GSM could co-exist in one unified 3G standard.

Backward compatibility will be important to the success of 3rd generation systems. When 3rd generation systems are initially deployed around 2002, there will be large groups of users on different digital cellular technologies. By year-end 2002, world-wide market projections show that GSM will have 279.4 million customers, CDMA

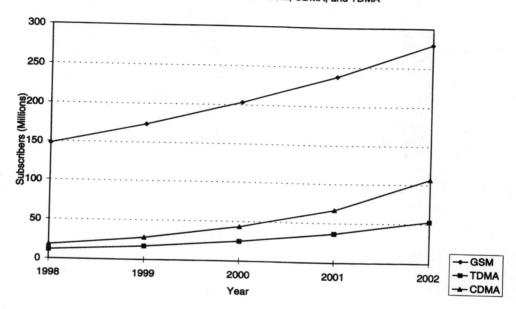

Figure 10.8, Subscriber Growth for Key Cellular Technologies
Source: Strategis Group

will have approximately 106.7 million subscribers and TDMA systems will have 52.8 million [4]. Figure 10.8 shows the projected subscriber growth for GSM, CDMA (IS-95) and TDMA (IS-136) between the year 1998 and 2002.

3rd Generation Projections

Projections show that the total number of 3G subscribers for 2006 will be 32.3 million [5]. The estimated system equipment cost per subscriber is $750 and the number of 3G handset sales should be 7.5 million units in 2006. During 2006, it is expected that 70 percent of the population to will be covered by some form of 3G system in 2006. The average subscriber unit cost should be approximately $400.

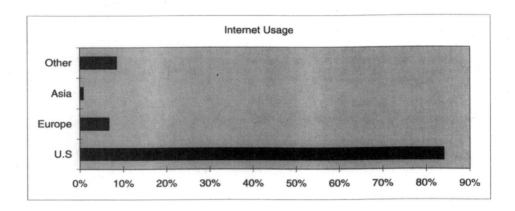

Figure 10.9, Internet Usage

Internet usage statistics can be used as a predictor of the need for 3G high speed services. The Georgia Institute of Technology's Graphics, Visualization and Usability 8th World Wide Web survey reveals interesting statistics about where the Internet is going and how its deployment in other countries is developing. During 1998, approximately 84 percent of all Internet users are located in the United States, while only 6.7 percent and 0.8 percent are located in Europe and Asia, respectively. Figure 10.9 shows amounts of Internet usage by country in 1998.

The success of the CDMA Smartphone, the Nokia Communicator, and the wireless phone manufacturing industry's rush to make competing smartphones in response, evidences consumer and business users' desire for Internet type capabilities on their phone. Further evidence is demonstrated by a recent Consumer Electronics Manufacturers Association (CEMA) survey, which places interest in a combination of traditional telephone and Internet access devices at 51 percent.

There are other signals occurring in the wireless industry that indicate the Internet-capable mobile phone trend is getting stronger. The sudden popularity of personal digital assistants (PDAs)— which is marked by the increased competition occurring between manufacturers in all segments of the information technology industry — from Microsoft and 3Com to Ericsson and Samsung — is proof that there is a need for convenient mobile computing. Another CEMA survey reveals that 58 percent of notebook PC owners would consider purchasing a hand-held PC instead of another notebook PC. With initial synergies in the global market pointing toward a highly favorable environment for Internet mobile phones, operators concerned about the lack of demand necessary to fuel 3G adoption should feel more assured.

Applications for 3rd Generation

The development of 3G applications is a question of what capabilities it can provide. This is already being evidenced in the United States thanks to increased landline connectivity speeds. As the Internet provides more applications that require high data speed, adoption by wireless devices such as 3G will follow. When email first became available in the late 1980;s, it was relatively slow to adopt. However, with increased PC and modem penetration, email became the Internet's first key successful application. The second successful application was world wide web browsing; hypertext markup language (HTML). Since HTML, the question has been what is the next big application. Some of the leading applications are broadcast audio (streaming), voice over the Internet and online purchasing. However, these are all relatively low bandwidth applications.

The applications for 3rd generation must require high bandwidth that only 3rd generation can supply. CEMA's Second Annual Multimedia PC survey shows that, with the rapid advent of faster modems (33.6 Kbps and 56 Kbps), consumers are reaching the Internet and accessing multimedia content much faster today than two years ago. Survey results show that only 13 percent of multimedia PC owners currently use a modem that transmits at 14.4 Kbps or less. Twenty-seven percent use a 28.8 Kbps modem, 28 percent use 33.6 Kbps, and 17 percent use 56 Kbps.

CEMA's Second Annual Multimedia Survey also revealed the increased use of audio technologies for the Internet. As of January 1998, Web browser plug-ins RealAudio and Shockwave commanded 68 percent and 36 percent market shares, respectively. Comparatively, in December of 1996, RealAudio had a 32 percent market share and Shockwave had a 13 percent market share.

With better technologies, computer telephony has also begun to take off. The 1998 MultiMedia Telecommunications Market Review and Forecast — a joint publication produced by the Telecommunications Industry Association (TIA) and the MultiMedia Telecommunications Association (MMTA) — estimate that the total spending on computer-telephone integration (CTI) equipment reached $1.3 billion in 1997, a 49 percent increase over 1996. The two associations' predict the CTI mar-

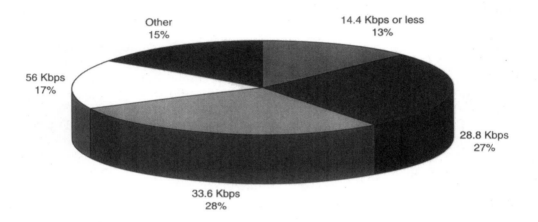

Figure 10.10, Percentage of Users' Modem Speed

ket will continue its phenomenal growth and is projected to reach $4.98 billion in 2001.

Of all the high bandwidth applications, video seems to be the most promising applications for 3G. To provide basic video service, bandwidth capabilities need to reach at least 144 kbp/s. As the bandwidth from 3G becomes available – alternative sources of high bandwidth will become available. These include high speed over the telephones lines (xDSL), satellite systems or cable modems. The applications developed for these transport providers is likely to drive the need for wireless applications that also require high data transfer rates.

Competition to 3rd Generation

There are several new high speed data transport systems that have been introduced or will be introduced to the market soon. These digital subscriber lines (DSL), cable modems, wireless cable and satellite systems.

The first relatively high speed data services was provided through integrated subscriber digital line-based (ISDN) services. ISDN provides data transfer rates to homes and businesses at 144 kbp/s. ISDN was commercially introduced in the mid 1990's and is somewhat popular in Europe. Unfortunately, due to the advances in analog modem technology and discouraging price plans offered by telephone companies, ISDN has not been very successful in North America. However, through the use of advanced signal processing technology and low cost integrated circuits, it is now possible to send high speed data service through standard telephone lines at data rates exceeding 6 Mbp/s. This technology is called digital subscriber line (DSL). There are various versions of DSL: HDSL, ADSL, VDSL and others. This has resulted in the common abbreviation for DSL technology as xDSL. If telephone companies offer xDSL promote and offer xDSL service at competitive pricing, it could replace some of the high cost leased lines.

The cable industry is starting to target the telecommunications market through the introduction of cable modems. High-speed cable modems are available in some metropolitan markets although their deployment has been slower than expected. This is likely due to the targeted low revenue residential customer base, which are primarily consumers and not the more lucrative business market.

A recent industry to develop that is competing is wireless cable. The wireless cable industry has evolved to provide high speed data services (up to 32 Mbp/s) to end customers. There are three key versions of wireless cable; Multichannel Multipoint Distribution Service (MMDS), Local Multipoint Distribution Service (LMDS) and Multichannel Video Distribution System (MVDS). MMDS was the first wireless cable system to be introduced. It operates at approximately 2.6 GHz and has a range that approaches 50 miles. LMDS operates at 28 GHz and MVDS operates near 38 GHz. The amount of bandwidth these systems can offer is significant (over 1 Gbp/s for LMDS) and the frequencies for these systems can be re-used similar to a cellular system. Analysts believe the high-speed Internet capabilities of these wireless systems — and not wireless cable video — will be the initial selling point for their systems.

Satellite systems are converting to high capacity digital transmission service. Portions of this high speed digital service can be used by Internet and other data users. Current video competitor DirecTV operator Hughes Network Systems is already marketing a broadband solution dubbed DirecPC. The turbo version of the technology provides users downlink speeds of more than 400 kbps. The new proposed low earth orbit (LEO) satellite systems including TeleDesic and Celestra propose high speed data links available globally.

An additional factor to consider for the deployment of new systems is the World Trade Organization (WTO) General Services Agreement that deregulates telecommunications foreign investment rules for telecommunications companies. Foreign technologies will be able to invest up to 49 percent in different countries with each country maintaining a different implementation schedule.

The majority of signatory developing countries will be in accordance with the Agreement by 2006. This is about the same time 3G systems will reach full steam in market strength. As a technology in the competitive environment, 3G offers carriers voice, data and compressed digital video services coupled with a quick rollout. This can allow 3G to become the ultimate wireless local loop system. If the radio spectrum is allocated, new operators can adopt 3G and not only compete with incumbent operators, but in some instances will also have a technological edge over the incumbent thanks to 3G's broadband capabilities.

References:

1. Nils Rydbeck, Sandeep Chennakeshu, Paul Dent, Amer Hassan, "Mobile-Satellite Systems: A Perspective on Technology and Trends", Proceedings of IEEE Vehicular Technology Conference, April-May 1996, Atlanta, Georgia, USA.
2. ibid.
3. ibid
4. Strategis Group, "World Cellular and PCS Markets study, Washington DC, 1997.
5. ibid.

Appendix I - Acronyms

A	interface between MSC and BSC
ACCOLC	Access Overload Class
ACD	automatic call distributor
ACIPR	adjacent carrier interference protection ratio
ACK	Acknowledge
ADC	analog to digital converter
ADPCM	Adaptive Differential Pulse Code Modulation
AEC	Acoustic Echo Control
AGC	Automatic gain control
AGCH	Access Grant Channel
AIN	Advanced Intelligent Network
AM	(1) Amplitude modulation; (2) ante meridian, before noon.
AMA	Automatic Message Accounting
AMPS	Advanced Mobile Phone System
ANI	Automatic Number Identifier
ANSI	American National Standards Institute
ARQ	Automatic ReQuest to retransmit
ARTS	Advanced Radio Technology Subcommittee
ASCII	American standard code for information interchange
ASIC	Application Specific Integrated Circuit
ASK	Amplitude Shift Keying
ATM	Asynchronous Transfer Mode
AuC	Authentication Center (data base). Associated with HLR.
B-CDMA	Broadband CDMA
B/I	Busy Idle Bit
BCC	Base Station Color Code
BCCH	Broadcast control channel
BCH	Broadcast channel
BER	Bit error rate
BIS	Busy Idle Status
BOC	Bell Operating Company
bps	bits per second
BPSK	Binary Phase Shift Keying
BS	Base station
BSC	Base station controller
BSS	Base station Sub-system
BT	Base Transceiver
BTA	Basic (rural, suburban) Trading Area
BTS	Base transceiver station (or system)
C/(I+n)	Ratio of carrier to the sum of interference and noise
C/I	Carrier to interference ratio

C/N	Carrier to Noise Ratio
CAI	Common Air Interface
CAP	Competitive Access Provider
CCF	Conditional Call Forwarding
CCH	Common Control Channel
CCIR	Committee on International Radiocommunications
CCITT	Consulatative Committee on International Telegraphy and Telephony
CCS	One hundred call seconds (traffic unit)
CCS7	Common channel signaling system Number 7
CDG	CDMA Development Group
CDMA	Code Division Multiple Access
CDPD	Cellular digital packet data
CDR	Call Detail Recording
CDVCC	Coded digital voice color code
CEPT	Conférence Européenne (des Administrations) des Postes et des Télécommunications
CFB	Call Forwarding on mobile subscriber Busy supplementary service
CFNA	Call Forwarding-No Answer
CFU	Call Forwarding Unconditional supplementary service
CGSA	Cellular Geographic Service Area
CIR	Carrier to Interference Ratio
CLEC	Competitive Local Exchange Carrier
CLI	Calling Line Identity
CLID	Calling line identification; caller ID.
CLIP	Calling Line Identification Presentation supplementary service
CLIR	Calling Line Identification Restriction supplementary service
CMAC	Control mobile attenuation code
CODEC	Coder-decoder
CPE	Customer Premises Equipment
CPP	Calling Party Pays
CRC	Cyclic redundancy code/check
CSMA	Carrier Sense Multiple Access
CSPDN	Circuit Switched Public Data Network
CT(n)	Cordless Telephony (nth generation)
CTI	Computer Telephony Integration
CTIA	Cellular Telecommunications Industry Association
CTRC	Canadian Television and Radio Commission (successor to DOC)
CWTA	Canadian Wireless Telecommunications Association
D/R	Distance to cell radius ratio
dB	Decibel
dBm	Decibel level relative to a 1 milliwatt reference level.

DCC	Digital Color Code
DCCH	Dedicated Control Channel
DCE	Data Communications Equiptment
DCS	Digital Cellular System
DID	Direct Inward Dialing
DRx	Discontinuous Reception(mechanism)
DSI	Digital (or dynamic) speech interpolation
DSL	Digital Subscriber Line
dsp	Digital signal processing
DTC	Digital traffic channel
DTE	Data Terminal Equipment
DTMF	Dual-Tone Multi-frequency (signaling)
DTX	discontinuous transmission
Ec/No	Ratio of energy per modulating bit to the noise spectral density
EIA	Electronics Industries Association
EIA-553	Industry standard for the AMPS cellular system
EO	End Office
ERP	Effective Radiated Power
ESN	Electronic Serial Number
ETSI	European Telecommunications Standards Institute
FCC	Federal Communications Commission
FDD	Frequency division duplex
FDM	Frequency Division Multiplexing
FDMA	Frequency Division Multiple Access
FDX	Full Duplex
FEC	Forward Error Correction
FER	(1-) Frame Error Rate (2-) Frame Erasure Rate
FEX	Foreign Exchange
FHMA	Frequency Hopping Multiple Access
FM	Frequency modulation
FOCC	Forward Analog Control Channel
FPLMTS	Future Public Land Mobile Telephone System
fps	Frames Per Second
FSK	Frequency Shift Keying
GEO	[1]-Geostationary Earth Orbit; [2]-Geosynchronous Earth Orbiting
GHz	Gigahertz, a thousand million cycles per second.
GPS	Global positioning system
GSA	Geographical Service Area
GSM	Global System for Mobile communication (formerly Groupe Spècial Mobile)
HLR	Home location register
HTML	Hyper-Text Markup Language

Hz	one frequency unit hertz, equal to one cycle per second
IPR	Intellectual Property Rights
IS	Interim Standard
IS-136	Interim standard 136 for North American TDMA cellular system
IS-41	Interim standard 41 for North American inter-switch signaling
IS-95	Interim standard for CDMA cellular service.
ISDN	Integrated Services Digital Network
ISO	International Standards Organization
ISP	Internet Service Provider
ITU	International Telecommunication Union
IVCD	Initial Voice Channel Designation
IVR	Interactive Voice Response
IWF	Inter-Working Function
IXC	Interexchange Carrier
kbps	kilobits per second, a thousand bits in one second
LAN	Local Area Network
LAP	Link Access Protocol
LEC	Local Exchange Carrier
LEO	Low Earth Orbit
MEO	Medium/Middle Earth Orbit
MFJ	Modified Final Judgment
MHz	Megahertz, a million cycles per second
MIN	Mobile Identification Number
MIPS	Million Instructions Per Second
MODEM	Modulator-demodulator
ms	(1-) millisecond(s); (2-) mobile station, mobile set
MSA	Metropolitan Statistical Area
MSC	Mobile service Switching Center
MSS	Mobile Satellite Service
MTA	Major/Metropolitan Trading Area
MTSO	Mobile Telephone Switching Office
MTX	Mobile Telephone Exchange
NACN	North American Cellular Network
NAM	Number Assignment Module
NAMPS	Narrowband Advanced Mobile Phone Service
NANP	North American Numbering Plan
NMT	Nordic Mobile Telephone, as in NMT-450, NMT-900
NPA	Numbering Plan Area (area codes)
NPRM	Notice of Proposed Rule Making
NSS	Network Subsystem
OA&M	Operations, Administration and Maintenance
OACSU	Off Air Call Set Up

OQPSK	Offset Quadrature Phase-Shift Keying
OSS	Operator Services System or Operational Support System
PABX	Private Automatic Branch Exchange
PAM	Pulse Amplitude Modulation
PCH	Paging Channel (GSM related systems)
PCIA	Personal Communications Industry Association
PCN	Personal Communication Network
PCS	Personal Communication Service
PDA	Personal Digital Assistant
PIN	Personal Identification Number
PLL	Phase Locked Loop
PLMN	Public Land Mobile Network
PN-PRBS	Pseudo-noise, Pseudo-random binary string/stream
POTS	Plain Old Telephone Service
PSC	Public Service Commission
PSK	Phase Shift Keying
PTT	Postal Telephone and Telegraph
PUC	Public Utilities Commission
QAM	Quadrature Amplitude Modulation
QPSK	Quadrature (four angle) phase shift keying (type of modulation)
RACE	Research (and Development) of Advanced Communication (Technologies) in Europe
RBOC	Regional Bell Operating Company
RF	Radio frequency
RSA	Reliable Service Area (for paging) or Rural Service Area (for cellu-lar)
RSSI	Received Signal Strength Indicator (or Indication)
RX	(Radio) Receiver
S/I	Signal to interference ratio
S/N	Signal to noise ratio
SAT	Supervisory audio tone (analog cellular systems)
SCM	Station Class Mark
SDMA	Spatial Division Multiple Access
SID	System Identification
SIM	Subscriber identity module
SINAD	Signal+Noise and Distortion
SNMP	System Network Management Protocol
SNR	Signal to Noise Ratio
SONET	Synchronous Optical Network
SS7	alternate abbreviation for CCS7 (signaling system 7)
ST	Signaling Tone
STP	Signal Transfer Point

SWR	Standing Wave Ratio
T-Carrier	Trunk Carrier
TACS	Total Access Communications System
TAPI	Telephony Application Programming Interface
TASI	Time assignment speech interpolation (see DSI)
TCP/IP	Transmission control protocol/Internet protocol
TDD	Time division duplex
TDM	Time Division Multiplexing
TDMA	Time Division Multiple Access
TIA	Telecommunications Industry Association
TLDN	Temporary Local Directory Number
TMN	Telecommunication Management Network
TMSI	Temporary Mobile Service Identity
TRA	Telecommunications Resellers Association
TRX	(Radio) Transceiver
TSAPI	Telephony Services Application Programming Interface
TSI	Time Slot Interchange
Um	radio link interface between MS and BS
UPR	User Performance Requirements
URL	Universal Service Locator
USTA	United States Telephone Association
VCGS	Voice Group Call Services
VLSI	Very Large Scale Integrated Circuits
WARC	World Administration of Radio Conference
WB-CDMA	Wide Band Code Division Multiple Access
WDF	Wireless Data Forum
WLL	Wireless Local Loop
WOTS	Wireless office telephone/Telecommunication system
WPBX	Wireless PBX
WWW	World Wide Web

Appendix 2 - CDMA Equipment Manufacturers

The following is a list of manufacturers that are involved in developing or providing equipment for IS-95 CDMA systems. This list was provided by the CDMA development group (CDG). The CDG is can be contacted at 1-704-540-1030, web: www.CDG.org and they are located at 575 Anton Blvd, Suite 530, Costa Mesa, CA 92626 USA.

Handset Manufacturers

Acer Peripherals, Inc.
ALPS Electric (USA), Inc.
Audiovox Cellular Communications
Denso International America
Fujitsu Network Communication, Inc.
Hyundai Electronics Industries Co., Ltd.
Kyocera Corporation
LG Information and Communications, Ltd.
Lucent Technologies, Inc.
Motorola, Inc.
NEC America. Inc.
Nokia Corporation
OKI Telecom
Philips Consumer Communications
QUALCOMM, Incorporated
Samsung Electronics Co., Ltd.
Sharp Laboratories of America
Sony Electronics

Supporting Equipment Companies

3Com
ADC NewNet Inc.
ALCAtel
Aldiscon Enterprises Inc.
ALLGON Telecom Ltd.
Certicom
Compaq Computer Corporation
DSP Communications, Inc.
ETRI
Fujant, Inc.
KSI, Inc.

LSI Logic
Metawave Communications
Mobile Systems International
MMCD/Panasonic
NEC Electronics
RADWIN
SCALA
Sema Group
Siemens Microelectronics, Inc.
SignalSoft Corporation
SnapTrack, Inc.
Sun Microsystems Computer Corporation
Synacom Technology, Inc.
Telular Corporation
Texas Instruments
TruePosition
Unwired Planet
VLSI Technology, Inc.
Wireless Facilities, Inc.

Test Equipment Manufacturers

Anritsu
Comarco Wireless Technologies, Inc.
Grayson Wireless
Hewlett Packard
IFR Systems, Inc.
Innovative Global Solution, Inc
Noise Com, Inc.
Racal Instruments
SAFCO Technologies
Tektronix, Inc.
Telecom Analysis Systems
Wavetek Corporation

Network Equipment Manufacturers:

Ericsson
Hitachi Telecom (USA), Inc.
Hughes Network Systems, Inc.
Hyundai Electronics Industries Co., Ltd.
LG Information and Communications, Ltd.
Lucent Technologies, Inc.
Motorola CIG
NEC do BRASIL, S.A.
Nortel Networks
QUALCOMM Incorporated
Repeater Technologies
Samsung Electronics Co., Ltd.

Appendix 3 - CDMA World Listings

The following is a list of wireless carriers that provide or in the early test stages of providing IS-95 CDMA service. This list was provided with the assistance of the CDMA development group (CDG). The CDG is can be contacted at 1-704-540-1030 and they are located at 575 Anton Blvd, Suite 530, Costa Mesa, CA 92626 USA.

Country	Carrier	System
Argentina	Compania de Radiocommunicaciones Moviles	Cellular
Australia	AAPT	PCS/PCN
Australia	Hutchison Telecom	PCS/PCN
Australia	OzPhone (consortium)	
Australia	Telstra	Cellular
Bangladesh	Bangladesh Rural Telecom Authority	WLL
Bangladesh	pacific Bangladesh Telecom Ltd (PBTL)	Cellular
Brazil	Centais Telefonicas de Ribeirao Preto (CETERP)	Cellular
Brazil	GlobalTelecom do Brasil	Cellular
Brazil	Telebahia	WLL
Brazil	Telebahia	Cellular
Brazil	Telebrasilia	WLL
Brazil	Telecommuncacoes do Rio de Janeiro SA	Cellular
Brazil	Telecommunicacaoes de Minas Gerais (TELEMIG)	WLL
Brazil	Telecommunicacoes de Mines Gerais (TELEMIG)	WLL
Brazil	Telesp	Cellular
Canada	BCTel	WLL
Canada	BCTel	Cellular
Canada	BCTel	PCS/PCN
Canada	Bell Mobility	WLL
Canada	Bell Mobility	PCS/PCN
Canada	Bell Mobility	Cellular
Canada	Clearnet Communications	PCS/PCN
Canada	MT&T Mobility	PCS/PCN
Canada	MTS Mobility	PCS/PCN
Canada	Telus Mobility	PCS/PCN
Chile	Chilesat Telefonia Personal SA	PCS/PCN
China	Beijing Telecommunication Administration (BTA)	WLL
China	China MPT	WLL
China	China United Telecommunications (CUT)	
China	China's Great Wall	Cellular
China	Communications	Cellular
China	Shanghai Great Wall Mobile Communication Corp.	Cellular
China	Shanghai PTA	WLL

Country	Carrier	System
China	Shenda Telephone Company	WLL
Congo	Telecel International	Cellular
Dem. Republic of Congo	African Telecommunications Inc. (AfriTel)	WLL
Dominican Republic	Compania Dominicana de Telefons	PCS/PCN
Dominican Republic	TRICOM	Cellular
Egypt	ARENTO	WLL
Egypt	Telcom Egypt	WLL
Germany	T-Mobile	Cellular
Guadeloupe, French West Indies	France Telecom	PCS/PCN
Guatemala	Telecomunicaciones de Guatemala (Guatel)	PCS/PCN
Guatemala	Telecomunicaciones de Guatemale (Guatel)	WLL
Hong Kong	Hutchison Telecom	Cellular
India	Bharti Telenet, Ltd.	WLL
India	Mahangar Telephone Nigam Ltd. (MTNL)/DOT	WLL
India	Shyam Telelink Ltd.	WLL
India	Tala Teleservices Ltd./BCI	WLL
Indonesia	PT Komselindo	Cellular
Indonesia	PT Telekomunikasi (Telkom)	WLL
Israel	Pele-Phone Cellular Communications	Cellular
Japan	Cellular Telephone Co. Group (DDI)	Cellular
Japan	IDO	Cellular
Korea	Hansol PCS	PCS/PCN
Korea	Korea Telecom Freetel	PCS/PCN
Korea	LG Information & Communications	PCS/PCN
Korea	Shinsegi Telecom Incorporated	Cellular
Korea	SK Telecom	Cellular
Kuwait	Ministry of Communications	WLL
Mauritius	Mauritius Telecom	WLL
Mexico	Baja Celular Mexicans (BAJACEL)	Cellular
Mexico	Celular de Telephonica (CEDETEL)	Cellular
Mexico	Grupo Lusacell (COMCEL, PORTACEL, TELECOM, & SOS)	PCS/PCN
Mexico	Grupo Lusacell (COMCEL, PORTACEL, TELECOM, & SOS)	Cellular
Mexico	Movitel del Noroeste (MOVITEL)	Cellular

Country	Carrier	System
Mexico	Pegaso Telecommunications	PCS/PCN
Mexico	Sistemas Profesionales de communication (SPC)	PCS/PCN
Mexico	Telefonica Celular del Norte (NORCEL)	Cellular
Mongolia	Mobicom	WLL
Nigeria	Intercellular Nigeria Ltd.	WLL
Nigeria	Nigerian Starcomms Ltd.	WLL
Peru	Telefonica del Peru	Cellular
Philippines	Express Telecommunications Co.	Cellular
Philippines	Pilipino Telephone Corp. (Piltel)	Cellular
Philippines	PT&T	WLL
Poland	Telecom Poland	WLL
Poland	TPSA (Telekomunikacja Polska SA)	WLL
Puerto Rico	Centennial	
Russia	Chelyabinsk Svyazinform	Cellular
Russia	Chelyabinsk Svyazinform &S. Urals Cellular Tel. Co	WLL
Russia	Electrosviaz	WLL
Russia	IV Telecom	Cellular
Russia	JSC Elecrtosviaz-Rostov	WLL
Russia	JSC Personal Communications	WLL
Russia	Kodotel	WLL
Russia	Kubtelcom	WLL
Russia	MTU-Inform/Pcomm	WLL
Russia	Tech Info Bellum	WLL
Singapore	MI (MobileOne)	PCS/PCN
Thailand	Tawan/Communications Authority of Thailand	Cellular
Ukraine	Telesystems of Ukraine	WLL
United Kingdom	Vodaphone	
United States	AirTouch Communications Inc.	Cellular
United States	Altel	Cellular
United States	Ameritech Cellular Services	Cellular
United States	Bell Atlantic Mobile	Cellular
United States	Frontier Cellular	Cellular
United States	GTE Wireless	Cellular
United States	Sprint PCS	PCS/PCN
Venezuela	Telcel	Cellular
Yemen	Public Telecommunications Corporation	WLL
Zambia	Telecel International	Cellular

Index